# THE DEFINITIVE GUIDE TO WAREHOUSING

# THE DEFINITIVE GUIDE TO WAREHOUSING

## MANAGING THE STORAGE AND HANDLING OF MATERIALS AND PRODUCTS IN THE SUPPLY CHAIN

Council of Supply Chain Management Professionals

Scott B. Keller and Brian C. Keller

Vice President, Publisher: Tim Moore
Associate Publisher and Director of Marketing: Amy Neidlinger
Executive Editor: Jeanne Glasser Levine
Consulting Editor: Chad Autry
Operations Specialist: Jodi Kemper
Cover Designer: Chuti Prasertsith
Managing Editor: Kristy Hart
Project Editor: Deadline Driven Publishing
Copy Editor: Apostrophe Editing Services
Proofreader: Apostrophe Editing Services
Indexer: Angie Martin
Compositor: Bronkella Publishing
Manufacturing Buyer: Dan Uhrig

For information about buying this title in bulk quantities, or for special sales opportunities (which may include electronic versions; custom cover designs; and content particular to your business, training goals, marketing focus, or branding interests), please contact our corporate sales department at corpsales@pearsoned.com or (800) 382-3419.

For government sales inquiries, please contact governmentsales@pearsoned.com.

For questions about sales outside the U.S., please contact international@pearsoned.com.

ISBN-10: 0-13-344890-8
ISBN-13: 978-0-13-344890-0

Pearson Education LTD.
Pearson Education Australia PTY, Limited.
Pearson Education Singapore, Pte. Ltd.
Pearson Education Asia, Ltd.
Pearson Education Canada, Ltd.
Pearson Educación de Mexico, S.A. de C.V.
Pearson Education—Japan
Pearson Education Malaysia, Pte. Ltd.

Library of Congress Control Number: 2013952808

*Dedicated to Karen C. Keller—Mother, friend, and family logistician.*

# CONTENTS

# ACKNOWLEDGMENTS

We are grateful to Kathryn Cordeiro for her graphic and research support.

# ABOUT THE AUTHORS

**Scott B. Keller** is a professor of logistics and marketing at the University of West Florida. He received his Ph.D. from the University of Arkansas and has been on faculty at Penn State and Michigan State. His research interests include issues in personnel development and performance, and the development of market-oriented cultures within logistics operations. He has conducted research for numerous corporations, and his work has appeared in leading logistics journals. He is the co-editor of the *International Journal of Logistics Management,* an associate editor of the *Journal of Business Logistics* and a member of the Council of Supply Chain Management Professionals. His managerial experience is in warehousing, motor carrier operations, and ocean freight terminal operations.

**Brian Keller** became an independent consultant in 2006. In this capacity, he has supported commercial industry companies as well as Government entities including the Office of the Deputy Under Secretary of Defense for Innovation & Technology Transition, the Defense Advanced Research Projects Agency (DARPA), and the Defense Science Board. Previously, Keller was chairman and president of GMA Cover Corporation, a multinational company that designed, manufactured, and supported signature management products including the Ultra Lightweight Camouflage Net System (ULCANS). During Keller's tenure, GMA won the Department of the Army competitive procurement for a $1.7B ULCANS production contract. Prior to GMA, Keller was a vice president for Stewart & Stevenson (now part of BAE) where he was responsible for the Family of Medium Tactical Vehicle (FMTV) A1R program including the successful award of the $4B rebuy production contract. Keller completed a 21-year military career as a logistician, Lieutenant Colonel, and the Army Product Manager for Field Support Systems. He is an alumni of the Harvard Business School, received a Bachelor of Science degree from the United States Military Academy at West Point, an MBA degree from the Florida Institute of Technology, and an MS degree in industrial engineering from the University of Alabama.

# 1

# WAREHOUSING'S ROLE IN THE SUPPLY CHAIN

## Introduction

This chapter explores warehousing's expanded role in customer operations and supply chain management. You learn about historical and current examples of warehouse support to manufacturing, purchasing, and various economies of operations. This chapter discusses competitive supply chain strategies, providing examples of value-added services that warehouses can now provide. With the expansion from a one-dimensional storage repository to a main element of customer supply chains, the warehouse is now expected to contribute to the overall client business objectives and contribute to cost reductions.

## Warehousing's Role in the Supply Chain

Warehousing played a role in the storage and exchange of goods for centuries. Long-term storage to provide product for future consumption has been a utility of warehousing both past and present. Transit sheds, warehouses connected to a wharf, have facilitated the movement and storage of goods embarking or disembarking merchant and military vessels supplying domestic and world trade. Rail transportation set in motion the industrial era with the transport of agriculture commodities and livestock; warehousing was leveraged to store such cargo prior to processing and then distribute finished products traveling to other parts of North America.

Long-term storage and places to interchange products may have been enough utility prior to and during the initial stage of industrial development; however, U.S. involvement in World War II required the manufacturing of products to support military efforts. Increased manufacturing demanded more storage and organization of raw materials and parts, as well as more room for the stockpiling and strategic positioning of completed military products from ammunition and vehicles, to food stores. Figure 1-1 depicts a high-cube military storage warehouse. Warehousing became more of a strategic function in the chain of supplying the U.S. military and its allies.

Figure 1-1    High-cube military storage warehouse

### Army Warehouses in World War II

During World War II the U.S. Army established supply warehouses in the state of Washington. These depots played critical roles in supplying the war effort in the Pacific. The depots warehoused large quantities of material. The warehouses in Washington delivered goods through the ports to support the war in the Aleutian Islands. They also supported the war in the Pacific by shipping critical equipment and supplies to Hawaii and beyond.

Engineering breakthroughs partially resulting from war efforts were adopted by industry post WW II. Although railroads provided dominance in freight transport prior to the World War II, motor carriers and eventually air carriers would surface as viable competition for freight transportation. Competitive changes, such as these, changed the face of warehousing. Now, a warehouse could receive a single truckload of product rather than a railcar load of product. Dynamics of unloading a tractor-trailer load compared to unloading a railcar are dramatically different and require differential planning for unloading and storage.

At the same time, developments were achieved in forklift handling equipment. Simple pallet jack capabilities were exceeded by higher reach forklifts enabling operators to build and manage freight in higher vertical storage buildings and reduce the fixed cost of engineering and fabricating the facility.

With the proliferation of computers, information exchange in the late 20th century became a game-changer in the way warehouses collected, transmitted, and utilized data and information within facilities and with warehouse customers. Perhaps computers came about in such good time to enable warehouse operators better control over the increasing variety of products demanded by consumers. Ever since the end of WW II, the United States realized a growing middle class society demanding a greater selection of products that required greater warehouse control. Product variations require greater skill in inventory control over that of managing a single commodity or a few finished goods items. Each unique product type requires a location in the warehouse that it cannot share with a different product type. Moreover, as market expansion spread, so too did the number of warehouses called upon to service the distant markets. Products to satisfy customer regions were dedicated to a specific market warehouse. Consequently, the aggregate inventory total for all market warehouses increased the investment in stock required to compete for market share.

# Traditional Roles of Warehousing

Although supply chains demand greater service value from warehouse operations, the basic economies of manufacturing, purchasing, and transportation must continue to be supported. Cost trade-offs, along with service expectations, must be evaluated to determine the role of the warehouse in supporting the traditional economies of scale.

## *Supporting Functional Economies of Scale*

Wide scope business strategies catering to broad-based clientele require large scale purchasing, production, and distribution. Achieving competitive scale demands operating efficiencies and economies supported by large scale warehousing of supplies and product. Economies of scale in purchasing, production, and transportation have long required warehouse support, and today, need continues for such warehouse support.

## Role in Supporting Economies of Manufacturing

Long manufacturing runs of single products create efficiencies in production processes, allocation of personnel, and capacity utilization of machinery and equipment. A manufacturer and marketer of a major brand of candy found that it would be financially feasible to operate a single production line for three flavors of a specific candy. To change from one product to another, the changeover process required that the machinery be completely disassembled, sterilized, and reassembled prior to running the next item on the master production schedule. Three days were required to complete the changeover, and the sterilization was critical because one of the three products included a nut ingredient. Sterilization reduced the threat of cross-contamination of products that could have devastating consequences if consumed by people with severe allergies toward nut products.

Plant supporting warehouses must add value in the supply chain by supporting long manufacturing runs to gain economies of production and reduce changeover needs. Single-item finished products produced in mass quantities must be stored and maintained for future demand.

## Role in Supporting Economies of Purchasing

Materials planners utilize the master production schedule and materials requirements plans to determine ordering needs for each material or component required to meet production plans. Planners and procurement personnel work together to evaluate material needs, lead times for receiving materials, and price-break concessions afforded to buyers for ordering in bulk quantities. All the components influence the need to receive and store materials and components for future production. Specifically, bulk purchase pricing may provide cost-savings per item that when purchased in great enough quantity it more than offsets the cost of storing and maintaining the materials.

Warehouse operators add value for manufactures, assembly operations, and consolidation points by receiving, storing, maintaining, picking, and shipping materials and components to support large volume purchase discounts. The need is further realized as variations in quality and lead times necessitate purchasing added safety stock to protect against such fluctuations.

## Role in Supporting Economies of Transportation

Similar to both manufacturing and purchasing economies, the better a carrier utilizes the full capacity and capability of its transportation equipment, the more efficient and cost-effective products are transported. Transportation cost per unit is reduced as a greater number of units are transported. Fixed costs are spread over the greater product amount being transported, and the variable costs do not necessarily increase one-for-one as another case of product is loaded onto a trailer and transported. Truckload (TL) business models are based on this premise, and truckload (LTL) and package carriers create bulk shipments by consolidating or bundling independent orders destined for a common ZIP code zone.

Costs associated with managing and holding greater levels of inventory in warehouse stock must be compared with the cost of transporting in large quantities to gain economies of transportation associated with reduced unit pricing. In many supply chains the transportation savings per case or item more than offsets the cost to warehouse additional product. Carriers can more efficiently utilize transportation equipment and offer discounts to shippers for helping carriers fill trailers. Warehouses add value by supporting large volume transportation needs.

# Demand for Contemporary Warehousing

Warehousing has been called upon by corporate to add value to supply chains while continuing to support traditional economies of scale and customer demand. As discussed, large storage warehouses are utilized to stockpile inventory that is produced, purchased, and transported in quantities large enough to gain competitive and cost-effective economies of manufacturing, procurement, and transportation. Such economies cannot be ignored by contemporary warehouse operators; however, additional factors must be considered when designing the strategy of the warehouse plan.

## *Anticipatory Inventory*

Many times products are produced in anticipation of demand and especially items that have a low cost associated with each unit. Brands associated with long historical demand data and with relatively predictable patterns (little unexpected variations in customer ordering quantities) may be prime candidates for producing in anticipation of the forecasted demand. Items with well-established demand patterns, low cost of goods sold, and minimal handling requirements would be kept in stock at levels to meet ordering and service requirements of customers. Although all inventory represents value in terms of dollars, items such as canned vegetables that have relatively steady base demand patterns, strong historical demand data for adjusting forecasts based on other relevant factors, ordered in case and/or pallet quantities, and require little value-added within the warehouse are potential items for anticipatory inventory.

## *Seasonal Stocking*

Red and white, and sometimes varied in colored, candy canes sold and consumed during the December holiday season represent an extreme case of seasonal stock. Manufacturers of the candy begin production and stock piling inventory well before orders are shipped to wholesalers and retailers. Historically, a southern U.S. candy maker would level production of the item by producing candy canes months in advance of demand so that labor and production machinery could more efficiently be utilized. Production strategies like this helped to reduce costs associated with overtime and running equipment near maximum capacity, thereby, risking an equipment breakdown. Producing well in advance of the season also allowed the candy maker to adjust production plans as the season approached. Compared to many other consumer products, candy canes are relatively low in cost per unit, require little handling without the cost of palletizing materials, may be stacked in high-bay storage, and are less susceptible to theft. As such, warehouse costs are more than offset by the reduction in production and labor costs.

## Balances Supply with Demand

It is infeasible to expect all customers to possess the capacity to order and receive full truckloads or even full pallet quantities of single items. Moreover, not all have the capability to store or the equipment to receive in large quantities, be it single items or mixed pallets. Warehouses offer storage to support production economies while also allowing customers the ability to order in lesser quantities and more often. Product assortment is available to customers so that they are not forced to receive and hold large quantities of single items in stock. In addition, warehouses receive products from various producers and offer a single point of interchange with the customer for distributing multiple items from multiple manufacturers. This minimizes the exchange points necessary between producers and their many customers.

## Protection Against Uncertainty in Demand and Lead Time

As previously discussed, seasonality may be a factor in the increase of demand for products sold and consumed during holidays or other seasons. Short-term marketing and sales promotions designed to stimulate customer purchases also must be considered in determining future demand while changes in business cycles and product life cycle trends may influence longer-term demand patterns for some products. Various influences on demand must be identified and taken into consideration when planning production; otherwise, left unknown, the factors may create an uncertainty in the quantity and assortment ordered by customers. Manufacturers will have to rush special production and carriers will have to expedite shipments; all adding cost to the supply chain while risking the loss of sales due to a product shortage when customers demand.

Warehouse inventory is compiled in anticipation of forecasted future demand. In addition, safety stock includes inventory on-hand to protect against any unknown influences that stimulate demand beyond the level forecasted. Under such conditions, warehouses are utilized to position and maintain stock in strategic locations where uncertainty exists and forecast accuracy is low.

In a similar manner, carrier on-time transit and delivery may fluctuate due to unforeseen circumstances or in extreme cases on-going poor quality of on-time delivery service. Marketers wanting high levels of in-stock availability will, in this case, hold a level of safety stock above the forecast to meet demand even if carrier deliveries are delayed.

## Competitive Supply Chain Strategies

Beyond supporting traditional economies of production, purchasing, and transportation, modern-day warehousing must assist in achieving corporate strategies designed to compete based on low cost and differentiation through various time-based strategies. Michael Porter, Harvard Business School professor and leading expert on competitive business strategy, and others have long established these as two overarching corporate-level strategies.

## Low-Cost Strategy

Low-cost corporate strategies may require long-term storage of large quantities of product. This was shown to support economies in production, purchasing, and transportation. Warehouses offer intermediate stocking points so that manufacturers do not have to service each individual final customer location. This allows manufacturers to ship in larger quantities to regional facilities servicing multiple end customers. The longest distance from the manufacturer to the regional warehouse utilizes truckload carrier service, thus leaving the shortest final distance for the more costly, yet flexible, LTL services. Overall, the total cost of transportation would be reduced with the help of the location of the regional warehouse (decentralized warehousing).

## Time-Based Strategies

While walking through a warehouse, a customer service manager looked up and said, "Look at all that candy." The accounting manager replied, "Look at all that money!" Twenty-first century supply chains must reduce costs and increase service to maintain competitiveness. Warehouses must do the same, and in ways unlike in the past break the service versus cost trade-off. By designing and adopting time-based strategies, supply chains may reduce inventory in the system and improve service responsiveness for themselves and their clients.

Firms are constantly seeking ways to reduce the lead-time from customer order placement to customer receipt of product and all while reducing levels of inventory in the system. Warehouses must contribute by instituting processes that are flexible and responsive to individual client needs. This may entail a cross-dock strategy, whereby, multiple shipments or items are received into the facility in bulk form and sorted according to final destination consignees. Orders for an individual consignee are then rebulked, loaded on an outbound trailer, and shipped to the destination without ever having been entered into storage.

Cross-docking and other time-based distribution strategies can assist in reducing supply chain system inventory, improving inventory turnover in stocking warehouses, responding better to customer lead-time requirements, adjusting to demand fluctuations, and reducing distribution facility costs. Postponement is another product customization and distribution strategy used to support firm-level, time-based market strategies. Intermediate or final stages of product customization are postponed until actual demand is realized; at which point the product is finalized according to customer specifications. Items are held in a higher level general state within materials or finished goods warehouse inventory until orders are received from customers.

## Interface Between Supply Chain Partners

Warehouses occupy strategic positions between suppliers and customers. Oftentimes, warehouse operators are the last personnel to see and touch products before final delivery. As such, they are the final entity to inspect product quality, condition, and count, and verify documentation accuracy. During any time of receipt, putaway, storage, picking, or loading products are vulnerable to cost increases. It is the efficiency, accuracy, and overall customer orientation of the warehouse operator that ultimately influences final customer perception and reality of quality and cost.

Managers of warehouses and their employees, alike, must interface with clients and customers of clients. Therefore, warehouses must be seen and managed as supply chain partners. Their impact can mean the success or failure of supply chain relationships between marketers and the ultimate customers.

## Critical Customer Service Role of Warehousing

For an order fulfillment center, customer service's role in order processing encompasses receiving the completed order form via an electronic or a paper device. On-hand inventory is checked to verify that the amount of stock requested on the order is in the warehouse and available to fill that specific customer's order. Stock availability, thereby, becomes a critical component of customer service that is influenced by the warehouse/order fulfillment center. Figure 1-2 illustrates warehouse racks consisting of multiple stock-keeping units with a majority of the slot locations having less-than-pallet quantities of product. Varied products and reduced inventory levels create challenges for warehouse operators to hold the correct amount of each product to satisfy customer demand.

When a stockout occurs and the item is not available in inventory when ordered, a customer must wait for the product to be replenished or authorize a substitute product to replace the original item ordered. Substituting a case of cherry breakfast pastries for a case of blueberry that was originally ordered may be of little consequence to the customer. (This is an assumption to make the point.) However, some products may not have suitable substitutes and a stockout could influence the customer to source from a competing supplier one time or for all future orders. Warehouses are often measured on stockout frequency or the related fill rate percentage of cases ordered (case fill rate = cases shipped / total cases ordered). This too impacts the ratio of orders shipped complete compared to total number of orders also known as order fill rate.

Figure 1-2    Storage of varied stock-keeping-units (SKUs) in varied quantities

Frontline warehouse operations also influence the condition of the product upon shipping. Damaged product arriving at a customer's facility may be denied and the bill of lading or delivery receipt adjusted at the receiving dock and the invoice cut or a claim initiated to recover the value of cases damaged. Percentage of damaged cases can be tracked over time to indicate severity and frequency of the problem. The number and type of a claim can be recorded and evaluated to identify potential issues and trends pertaining to specific items, customers, or warehouse order picking personnel.

Overages, shortages, and damages (OS&D) cause issues that oftentimes adversely impact multiple partners within the supply chain. Take for instance an issue discovered by the customer service director for a warehouse that managed the southeastern U.S. product distribution for multiple manufacturers of major national household brands of consumable products. In an effort to improve the standing of the warehouse in the eyes of customers, the customer service director began conducting field visits to the receiving docks of customers. When walking into a small wholesaler, the director was greeted by an angry and frustrated owner. The owner showed the director a closet filled with empty boxes that he claimed arrived empty and concealed within the interior cases on pallets. The director and owner set out to discover the root cause of the concealed, empty cases. Assuming the pallets were full pallets of single items, it could be that the cases were empty when they were palletized at the end of the production line. A second possibility is that the cases were emptied by warehouse or carrier personnel anywhere along the distribution channel.

At the end of the supply chain, the wholesaler's receiving personnel came in early mornings to break down pallets of product that were delivered during the night to a secured fenced area of the receiving dock. After careful investigation, it was determined that the wholesaler's personnel were breaking down the pallets, emptying and taking the product out of some of the boxes, and then reconfiguring the cases on the pallets where the empty cases would be concealed among the full cases. The owner would come in an hour later to find the issue and naturally assumed the shipping warehouse or carrier was at fault.

Two more critical service factors influenced by warehouses include the lead time required to process and ship an order from the time the order is received and the consistency of that lead time. Greater lead-time requires added inventory in the system to fulfill orders during the time orders are processed. This refers to cycle stock. Moreover, as lead-times fluctuate additional units of inventory are necessary to satisfy customer demand during times when the lead-time increases. Safety stock is necessary to protect against such fluctuations in lead-times caused by inefficiencies in warehouse processes.

Today's supply chains more often require flexible processes and partners. By working closely together to communicate alterations in demand and service needs, warehouse clients and operators can formulate the best circumstances for building flexibility in the warehousing and distribution system.

## Light Manufacturing and Assembly

Partners subscribing to the supply chain concept continuously search for more efficient and economical means to reduce supply chain costs. Here is where warehouses can add value beyond tradition. For example, a third-party warehouse (neither the manufacturer nor the customer) was storing wiring harnesses for a major automobile assembly plant. A plant in Mexico's Maquiladora region along the U.S. Texas border performed the laborious task of running and securing the many wires along each harness. To increase the value that the warehouse provided its customer, the warehouse operator drafted a proposal to perform the wiring of harnesses in the warehouse that is more strategically situated nearer the U.S. automotive assembly plant. The warehouse reduced the cost of the light manufacturing of the harnesses while also reducing the cost associated with the transportation and transit time required from the Mexican plant. Figure 1-3 illustrates a warehouse operation adding value by assembling tires to wheels that are then shipped just-in-time to the production line for final assembly on automobiles.

Oftentimes, certain light manufacturing or assembly activities can be more efficiently and effectively conducted within a warehouse instead of within a complex manufacturing plant. Under such circumstances, forward-thinking warehouse operators can add value in the supply chain by removing some of the manufacturing burden from the plant. This is especially beneficial given the warehouse has the capability to perform such processes to a level of expected quality and lead time all while reducing the cost to do so.

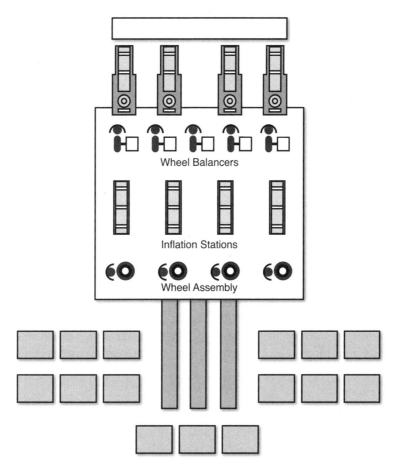

Figure 1-3    Warehouse value-added tire and wheel assembly

## Summary of Key Points

Warehousing's role in the supply chain has become more critical and at an escalating rate during the past two decades. Responsibilities of warehouse operators have evolved from maintaining long-term storage of materials and products to supporting economies of purchasing, production, and transportation to including light manufacturing and facilitating time-based supply chain strategies.

Warehouse operations contribute to the overall total cost of managing a supply chain, and as such, the trade-offs between warehousing costs and services to that of other critical functions of the firm must be evaluated. It is when warehousing contributes to reduced costs and improved service, flexibility, and responsiveness that warehouses become more valued to the organization and supply chain as a whole.

Value is provided through

- Storing product to fulfill customer demand and protect against uncertainties in demand and lead-time

- Providing customers with product assortment

- Postponing or delaying inventory commitment to form or location until demand is better known

- Achieving low total cost and improved lead-time through consolidating multiple orders

- Reducing lead-time through cross-docking

- Sequencing materials and components from multiple third-party logistics (3PLs) providers for time-based delivery to factory production lines

- Performing light manufacturing, assembly, and kitting

Most important, warehouses impact the receiving customer in many critical ways. Front-line warehouse personnel may be the final customer service defense in ensuring product accuracy, quantity, timing of shipment and delivery, accuracy of documentation, and overall product condition—all of which impact total cost and customer perception of the brand.

## Key Terms

- Anticipatory Inventory
- Bill of Lading
- Cost of Goods Sold
- Cross-Docking
- Cycle Stock
- Demand Patterns
- Distribution Channels
- Economies of Scale
- Fill Rate
- Fixed Cost
- Lead Time
- Less than Truckload (LTL)

- Mixed Pallets

- Overages/Shortages Damages (OS/D)

- Putaway

- Supply Chain

- Supply Chain Management

- Third-Party Logistics (3PL)

- Variable Cost

## Suggested Readings

Akerman, K. B. (1997, 2012), *Practical Handbook of Warehousing*, 4th ed., Chapter 1–2, Chapman and Hall, New York, NY.

Tompkins, J. A. and Smith, J. D., (1998, 2013) *The Warehouse Management Handbook*, 2nd ed., Chapters 1–5, Tompkins Press, Raleigh, NC.

Stock, J. R. and Lambert, D. (2001), *Strategic Logistics Management*, 4th ed., Chapter 10, McGraw-Hill, New York, NY.

# 2

# DISTRIBUTION CENTER CONCEPT

## Introduction

This chapter introduces the concept of the distribution center, differentiating it from the traditional, one-dimensional warehouse concept. The role of the distribution center in the flow of product through the supply chain is explored, as is cross-docking as a means of managing inbound and outbound freight. Sequencing, forecasting, and postponement strategies wrap up the chapter.

## Distribution Center Concept

Up to this point, the term warehouse has been used when describing storage all the way to light manufacturing. Historically, the term warehousing reflected the short-term, medium-term, and long-term storage of inventory at rest within a facility. In contrast, a distribution center takes on the role of facilitating the timed flow of inventory from shipper to customer to reduce the cost associated with storing high levels of inventory and improve service to meet the varied expectations of clients and their customers. Although the distribution center is a form of warehousing and stores product for future demand, the functionality and contribution of a distribution center exceeds that of a traditional storage warehouse.

## Facilitating Product Flow Through the Supply Chain

Distribution centers influence materials and product flow through supply chains by adopting many strategies; all of these should focus on providing for the success of users. Important considerations may include, for example, the need to accumulate inventory, break down inventory, provide product variety, sequence materials and parts replenishment, institute a postponement strategy, and cross dock freight for accelerating product flow through the supply chain.

## Accumulation, Sortation, Allocation, and Assortment

Supply chain partners closest to the customer often have little space for stockpiling inventory in large quantities. Most retail stores are designed to maximize the utilization of space for product sales. A distribution center can provide the point within the supply chain where inventory can be accumulated from multiple sources to meet future demand while supporting economies of manufacturing and transportation. Distribution centers respond to product sales by replenishing retail stock from the reserve accumulated inventory in the distribution center and allocating only the portion wanted by an individual retail store.

Distribution centers facilitate the exchange of product between sellers and buyers. As buyers become more knowledgeable and want greater variety and products tailored to their needs, distribution centers can provide product assortment in the mixture that customers want. One-to-one interactions between manufacturers and end customers are reduced as products are shipped in bulk quantities to distribution centers and thus allowing customers to source multiple SKUs from a single central point in the supply chain. The number of interactions is reduced, and the flow of product through the supply chain is more efficient and cost-effective.

Distribution centers also provide sortation services to separate products into groups related to customer desires for amount and quality grades of products. For example, a large retailer hired a 3PL to manage the reverse flow of distressed product from the retail store back to a distribution center that determines the final disposition of the items. Distressed product may be items that failed to sell, were damaged, or perhaps reached their predetermined "sell by" date.

The 3PL managing the reverse logistics for the retailer accumulated the distressed product and made enough bulk to gain economies of transportation back up the supply chain. When received in the returns facility the items were sorted, evaluated for function and quality, and a disposition was determined for each item. Items were then redistributed to another location for retail sale, refurbished, and sent to a discount store or exported to a market that could utilize the product, donated or destroyed if the product was beyond recovery. Common causes of product returns include:

- **Returns from the manufacturer**
  - Surplus of parts and materials
  - Poor quality inputs received
  - Excess production

- **Distribution returns**
    - Unsold stock
    - Product overage upon delivery
    - Product damaged in transit
- **Customer returns**
    - Defective products
    - Product recalls
    - Expired product

GENCO is a company that has expertise in managing returns for its clients. Moreover, it can assist in managing product recalls; traditionally a difficult process to manage, but one where timing may be critical to the brand image of the client.

## *Full Line Stocking Distribution Center*

Marketers of multiple product lines may rely on a distribution center to stock a full line of products that cover the offerings of the marketer for customers. Items manufactured at independent facilities are then made available through the distribution center allowing for customers to have full access to all the items in the product line.

## *Cross-Docking*

Less than truckload (LTL) and package freight operations require the utilization of cross-dock facilities that receive freight but do not hold freight in long-term inventory. Freight enters the facility, receives disposition, and exits for its final destination within hours. LTL motor carriers and parcel carriers operate multiple cross-dock facilities to manage the handling of orders from common origins and the consolidation of orders to common destinations. Cross-docking may also occur on a per order basis as inbound trucks are received into a distribution center and the unloaded freight is moved across the dock, bypassing storage, and reloaded on an awaiting outbound trailer.

## Bulking, Break-Bulking, and Flow-Through

While at its purest, a cross-dock provides value through quick break-bulk and make-bulk processes; a flow-through center adds value to the product as it moves through the distribution process. It may take hours or days to process, but like standard cross-docking, the product is already assigned a consignee and does not enter into traditional storage. A completely flow-through distribution facility distributes domestic and imported apparel directly to retail store locations in and around Birmingham, AL. Multiple stock keeping units (SKU) within multiple orders arrive at the facility. Carriers arrive with intermodal ocean containers, consolidated LTL shipments, and even bundled parcel packages. Consolidated and bundled (bulked) shipments contain multiple individual orders originating from common shipping points. When at the cross-dock facility, individual orders are deconsolidated (break-bulking) and reconsolidated (bulking) according to destination store locations. Within 24 hours all orders are received, processed, loaded, and shipped from the cross-dock leaving no freight in the facility until the next bulked shipments arrive.

Value is added through the process of break-bulking and bulking of orders. This enables the cost-efficient inbound and outbound transport of less-than-truckload (LTL) orders originating from multiple locations and destined for multiple locations. In addition, the facility performs inspections of items, repackages some items into smaller quantities for specific stores, and affixes retail price tags to some items.

## Sequencing

Large scale manufacturing may require delivery of materials, parts, and components directly to the assembly station along the production line. Actually, modern auto assembly plants like the Hyundai plant in Montgomery, AL, and the Mercedes plant near Tuscaloosa, AL, demand a minimal number of components be delivered on an on-going basis to keep the assembly line operating. Plants no longer want expensive materials and parts sitting idle and in large quantities at or near the production line. Inventory is pushed back up the supply chain.

Warehouse operators are acting as fourth-party logistics providers (4PL) when they manage the inbound organization of multiple 3PL suppliers and their associated materials. The special duties require the 4PL to work closely with planning and production to understand the master production schedule and inbound materials requirements so that each part may be sequenced to arrive precisely when the assembly plant requires and at the designated location along the production line.

## Postponement

Forecasting at its most accurate is by nature less than 100 percent perfectly matched with future demand. Unexpected buying behaviors of customers worsen the ability of marketers and producers to accurately predict demand. Marketers segment customers based on the differing wants, needs, and benefits sought of customer groupings. As customer segments increase, so does the potential to error when forecasting the demand of multiple segments for which substantial or accurate historical data may or may not exist.

For this reason, forecasters may find it beneficial to aggregate data pertaining to similar market segments and thus create an overall forecast for the base components of the products. In this manner an aggregate forecast is calculated and the "generic" components or products are produced and warehoused; however, the final customization of the product to meet the needs of individual segments is delayed or postponed until actual orders are received from the customer segments. Only then can final production of the product occur.

Under such conditions, warehouses must facilitate the postponement strategy by instituting processes that are highly responsive to the final production needs of manufacturing. In special cases, the warehouse may be called upon to perform the final customization of the product and ship it to the customer. This is the case of a warehouse merge-in-transit facility that assembles kits customized according to the details of each order. For example, a major retailer of knock-down furniture (disassembled) may utilize a merge-in-transit strategy to bring together shipments containing the wood patterned components from one supplier, the fasteners from a second supplier, and the instructions and marketing literature from a third supplier. Upon all three elements arriving at a central merge-in-transit cross-dock facility, they will be bundled and packaged together to complete the process prior to shipping to the customer.

## Summary of Key Points

This chapter introduces the concept of the distribution center and its mission to facilitate the timed flow of inventory from shipper to customer. In contrast to a warehouse that stores goods, and provides some additional value to the shipper's process, the distribution center reduces cost associated with storing material and provides an ability to enhance customer service. Strategies to manage the flow of product through a distribution center were discussed including cross-docking and various sorting strategies. The role of 4PLs in managing inbound materials and sequencing to meet customer requirements was discussed, as was the concept of postponement strategies.

## Key Terms

- Accumulation
- Allocation
- Assortment
- Break-bulk
- Cross-docking
- Deconsolidate
- Distressed Product
- Distribution Center
- Economies of Manufacturing

- Forecasting
- Fourth-Party Logistics Provider (4PL)
- Full Line Stocking
- Intermodal
- Knock-down Furniture
- Less-than-Truckload (LTL)
- Make-bulk
- Merge-in-transit
- Postponement
- Product Recall
- Reverse Logistics
- Sequencing
- SKU
- Sortation
- Third-party Logistics Provider (3PL)

## Suggested Readings

Baker, P. (2004), "Aligning Distribution Center Operations to Supply Chain Strategy," *International Journal of Logistics Management,* Vol. 15, No. 1: pp. 111–123.

Foulds, L.R. and Y. Luo. (2006), "Value-added services for sustainable third-party warehousing," *International Journal of Logistics Systems and Management,* Vol. 2, No. 2: 194–216.

Mukhopadhyay, S. K. and R. Setaputra, (2006), "The Role of 4PL as the reverse Logistics Integrator: Optimal Pricing and Return Policies, *International Journal of Physical Distribution and Logistics Management*, Vol. 36, No. 9: pp. 716–729.

Timo Ala-Risku, Mikko Kärkkäinen, and Jan Holmström, (2003) "Evaluating the Applicability of Merge-in-transit," *International Journal of Logistics Management*, Vol. 14 Iss: 2, pp.67–82.

Yang, K. K., J. Balakrishnan, and C. H. Cheng, (2010) "An analysis of Factors Affecting Cross Docking Operations," *Journal of Business Logistics,* Vol. 31, No. 1: 121–148.

# 3

# GENERAL WAREHOUSING AND DISTRIBUTION CENTER STRATEGIES

## Introduction

This chapter discusses the quantitative and qualitative distinguishing traits of public, contract, and private warehouses. User examples assist in determining what to consider when selecting the type of warehouse arrangement to employ. You also see a cost comparison example to convey the concept of the cost point of indifference. Finally, this chapter presents a case study exercise to emphasize the analysis aspects of selecting between warehouse alternatives. The case key is also provided.

## General Warehousing and Distribution Center Strategies

Three general categories of warehouses include public, contract, and private. Selecting one over the others requires knowledge of the role and strategy for the warehouse, skill of the intended operator, activity, volume, and costs associated with managing the warehouse. A warehouse may function as a general public warehouse for multiple clients while also operating as a contract warehouse for an individual client requiring more specialized product handling, value-adding processes, and dedicated space.

### Utility of Public Warehousing

The public warehouse category is flexible and many times least costly. Operators of public warehouses offer storage and general handling services for a variety of clients and products. Typical storage agreements may be renegotiated annually; however, some agreements allow for as little as a 30-day written notice of vacancy.

Public warehouses offer space as needed for clients that may have fluctuating needs for space and require basic storage all the way up to greater value-adding services. Although operating contracts should stipulate intended space requirements, many public warehouse operators can expand and contract space dedicated to a specific client as the client's volume and throughput demands. In the same manner, the warehouse gains efficiencies in allocation of labor across clients. Space and labor will be scheduled according to client needs, and when not utilized to facilitate one client, the space may be utilized for another. Efficiencies are gained because personnel may be allocated and shared across clients. Equipment may be assigned to one section of the warehouse and then the next shift can service completely different client freight. The capability for a public warehouse to shift available space, equipment, and personnel allows the provider to offer lower costs for servicing all clients. In addition, the fixed costs and general overhead cost factors associated with the entire facility can be allocated across clients, reducing costs for all users.

## When to Consider

Qualitative and quantitative factors must be taken into account when considering hiring a public warehouse operator to manage a firm's product storage and distribution. For example, a marketer of a new sugar substitute sweetener required storage in anticipation of the new product launch. The marketer produced and warehoused enough of the sweetener to test-market the product in designated regions. Under such circumstances the marketer would require short-term storage of the product that may or may not require future warehousing. Because there is no direct history of demand for the new product, on-going space needs may fluctuate greatly. No special handling was needed for the sweetener beyond the typical pallet and case quantity storage and movement.

Public warehousing may be used to service products with demand patterns that are predictable and to the contrary: unpredictable, unknown, minimal, or declining in demand. Operators may adjust spacing needs as the client's demand requires. Client agreements may entail a per case or unit price for receipt, storage, and handling. Total cost would vary according to the volume managed by the public warehouse. As throughput increases, total cost would increase. As throughput decreases, total cost would reduce, accordingly.

With today's focus on managing end-to-end supply chain costs, traditional storage facilities must gain competencies to offer clients more than just high-volume and low-cost storage. For some clients, efficient and low-cost storage may serve their purposes; however, some public warehouses have adapted to the ever-changing logistics environment by offering client services pertaining to cross-docking, merge-in-transit, postponement, and other time-based competitive offerings. Other services that may be offered could include freight consolidation and deconsolidation, freight payment, labeling, and other value-adding services that were not traditionally performed by early history public warehouses. Some primary considerations for utilizing public warehousing include

- Long-term or short-term storage.

- Increasing, decreasing or fluctuating product demand.

- Positioning low-volume inventory widely geographically.

- Value-added services but not to the extent that contract services would dictate greater dedication to facility, equipment, personnel, and special services and processes.

- Client possessing little to no warehouse operating skills.

- Client wanting to minimize capital investment and risk.

- Client wanting to minimize personnel cost and management.

Some may say that differences are less distinctive between a public and contract warehouse today. Market demands have been the encouraging factors leading public warehouses to offer services unlike ever before. Corporate demands have also driven the need for warehouses to offer greater value and contribution to meet the cost and service goals of supply chains or risk becoming obsolete. The most progressive public warehouses have well-defined market segments that they target based on the operating competencies they possess. One public warehouse went so far as to gain a Foreign Trade Zone (FTZ) designation over its facilities so that it could encourage importers to store their freight in the Zone and defer payment of Customs duties until the freight was sold. The freight had medium-term, high-cube storage needs, relabeling needs, and some of the product required dumping.

## Utility of Contract Warehousing

It is advisable to operate under a contract agreement pertaining to service, expectations, responsibilities, and pricing for public and contract warehousing. By the nature of contract warehousing, however, the client and provider have more at stake. Agreements are long term due to the specific nature of the services performed and the investment in space, equipment, and employee training to provide the specific needs—all of which are solely dedicated to servicing the client.

Marketers of a global brand of potato chips gained efficiencies by producing in long production runs of single items of potato chips. Full pallet quantities of individual SKUs were received by a contract warehouse and placed in reserve storage for future demand. A large-volume member based club-store retailer ordered pallet quantities of mixed flavors of the chips that would be displayed for customers on the pallet. Contract stipulations guaranteed a minimum level of throughput. The contract warehouse hired and trained the warehouse employees in the process of breaking single item pallets and configuring mixed pallets containing alternating flavors per layer of product on the pallet. The operation required space for handling the mixing process.

Marketers of a national brand of hair products shipped direct from plant in full pallet quantities of a single SKU of shampoo and a single SKU of conditioner. Promotional coupons were mailed to the contract warehouse. When a retailer placed an order, the shampoo and conditioner were depalletized, single bottles of each were combined, a coupon was affixed, and the new unit of product was processed through a shrink-wrap tunnel to produce a new SKU. Again, the labor was hired and trained, and the equipment was procured, set up, and utilized for that customer's specific needs. The additional commitment of the warehouse operator required that the client contract for a guaranteed throughput that would cover the added labor and equipment cost to perform the value-added service and provide the contract warehouse an acceptable profit level.

The examples illustrate the higher level of commitment to specified long-term services required of a contract warehouse over that of a public warehouse. Guaranteed throughput ensures a client of guaranteed space and the labor and equipment to perform the services. A client would be charged a rate comparable to the guaranteed throughput even if the level were not met.

## When to Consider

Manufacturers and marketers with product requiring special handling, kitting, or light manufacturing would consider a contract warehouse strategy. To capitalize on the economics of utilizing contract warehousing, a client would need to ensure a steady and increasing level of throughput. Consideration would also include the availability of contract warehouse space in the required geographic location, and it would certainly be of consideration should the manufacturer and marketer not have professional warehousing expertise to consider private warehousing.

Value-added services would require the contractor to obtain, train, and manage warehouse labor and equipment. For this reason, the contractor would require a guaranteed level of throughput to make the investment worth the contract. Although a warehouse can operate as a public and contract warehouse, the contract warehouse operation would in turn guarantee the labor, equipment, and service level stipulated within the contract. The resources would be dedicated more highly to managing the client's business than that provided by a public warehouse operator.

From a cost-perspective, a contract warehouse strategy would be considered over a public warehouse as the level of throughput is steady and potentially increasing. The fixed costs required to set up for the client's needs would reduce per unit as throughput increased. Clients would expect variable costs to be less than when utilizing a public warehouse strategy because the contracted processes, personnel, and equipment would be customized to meet the needs of the client.

Primary considerations for utilizing contract warehousing include

- Value-added services or special handling needed
- Significant and steady product throughput
- Postponement strategies
- Positioning inventory for single or multiple markets
- Client possessing little to no warehouse operating skills
- Client wanting to minimize capital investment and risk
- Client wanting to minimize personnel cost and management

When considering public or contracted warehousing arrangements, you must review the respective entities' industry certifications. Given that the client in either of these situations does not manage daily quality control, the possession of industry certifications can ensure that the selected public or contracted warehouse has met and maintains established operating standards.

## Utility of Private Warehousing

Situated in the Midwest, a manufacturer of high-end office furniture offers customers quality functionality, construction, and service. Superior distribution service is achieved through private warehousing and transportation of all products. Controlling all aspects of service from manufacturing of the finished product to distributing the product directly to customers affords the manufacturer total control over all factors influencing customer perception of the brand image.

Office furniture is constructed of various materials that may be chipped, scratched, torn, or dented during handling. The company hired and trained its people so well that front-line warehouse employees utilized traditional blanket wrapping and nesting of furniture utilized in the household goods moving business. Without foam packaging and rigid corrugated boxing, workers had to provide extra care when moving, handling, and loading product.

In this situation, the manufacturer believed it gained a competitive advantage by ensuring quality in all aspects of warehousing and delivering an order. Although there was little need for complex value-added service, there was a high need for superior handling and distribution of the furniture. The company did not want to risk entrusting such critical responsibilities to a public or contract warehouse operator.

Private warehousing strategies are necessary when manufacturers believe that service and cost to distribute can be accomplished in-house better than outsourcing the responsibilities to a 3PL warehouse.

## When to Consider

Investing in a private warehouse operation requires analysis to support the capital investment, but also requires evaluating the qualitative aspects of the corporation's needs and overall strategy. A manufacturer or marketer having abilities and competencies in managing product warehousing could consider operating its own product storage and distribution. Products requiring high levels of support services and warehousing quality control may be prime targets for private warehousing.

Because the operator bears all fixed and variable costs associated with owning and operating a private warehouse, the level of product throughput must be steady or increasing, and must be high enough to spread the costs across enough units to reduce the per unit cost below that of the contract warehouse option. For this reason, a private warehouse must have strengths in managing people and processes effectively and efficiently to keep service high and cost-to-serve low.

Primary considerations for utilizing private warehousing include

- High service quality needed.
- Significant and steady product throughput.
- User possesses strong warehouse operating skills.
- User service and quality more than offset capital investment and risk.
- User possesses strong skills to hire, train, and effectively manage personnel and associated costs.
- Special handling and value-added services needed.
- Postponement strategies.
- Positioning inventory for single or multiple markets.

## Practical Cost Differences

Public warehousing provides the marketer an option with minimal commitment to long-term storage costs including those associated with the building, labor, equipment, administration, and others. Cost to manage the product would be allocated to the number of units in, stored, and shipped. The assumption is that while both fixed and variable type costs would be allocated, the user cost to the marketer would be according to the client's activity and assessed per unit, and charges would begin as cases entered the facility.

On the other extreme, a private warehouse operation would bear all the fixed and variable costs associated with procuring the facility and equipment prior to any activity moving through the facility. Although the fixed costs would be significantly higher for the user of private warehousing compared to that of a user of public warehousing, the variable cost to the private warehouse user should be significantly below that charged by a public warehouse operator. Again, the assumption is that a private warehouse operator should possess the competencies to better control the costs to manage the processes for handling, storing, and manipulating the company's own products. User cost for a contracting warehouse services would also include fixed cost but to a lesser degree than that of a private warehouse. However the quality and control in managing the private warehouse should reduce the variable costs below that of both contract and public warehouse options. Total cost comparisons can assist users in selecting the most appropriate warehouse option when considering fixed and variable costs, and the level of anticipated product throughput.

## Cost Points of Indifference

Total cost points of indifference can be calculated given the warehouse user can estimate the fixed and variable costs associated with each alternative of public, contract, or private warehousing operations. Table 3-1 provides the cost figures for three warehouse alternatives.

Table 3-1   Warehouse Comparisons Using Points of Indifference

| Warehouse Comparisons | FC + VC(U) = FC + VC(U) | Point of Indifference |
|---|---|---|
| Public = Contract | $4.00(U) = $30,000 + $2.50(U) | 20,000 units |
| Contract = Private | $30,000 + $2.50(U) = $100,000 + 1.25(U) | 56,000 units |

Suppose that a client could obtain public warehousing for $4.00 for each unit managed through the warehouse. The actual cost would contain a portion of fixed and variable cost, but as a user the cost would vary directly with the number of cases throughput. Further suppose that a contract warehouse operator would charge $30,000 for start-up costs to manage warehouse operations specifically customized for an individual client's needs. Beyond the fixed cost investment, the user would be charged $2.50 per unit to manage product throughput. Under these cost assumptions the total cost to utilize a public warehouse would be lower until throughput equaled 20,000 units, at which point the total cost for public would be equivalent to that of contract ($80,000). It is important to recognize that the total cost point of indifference considers only the assumed fixed and variable costs but does not consider other qualitative-type factors discussed. Contract warehousing becomes of consideration as throughput exceeds 20,000 units. The same type of cost relationship can be seen between contract and private warehousing. Because of the high fixed cost capital investment associated with a private warehouse strategy, throughput would need to be above 56,000 units to be considered cost feasible compared with contracting.

As indicated in the hypothetical example, investment in private warehousing requires confidence in the growth of product moving through a distribution facility. Risk associated with the high fixed cost is too high given a product declining in demand. Certainly, public warehousing is the least risky option in this example.

## Summary of Key Points

Many factors must be assessed when selecting a warehouse strategy. Public, contract, and private warehousing alternatives must be evaluated from a qualitative and quantitative perspective.

Public warehousing is easily procured and equally easy to divest. Although the cost per unit may be higher than that for other alternatives, the reduced risks associated with using a public warehouse offset the cost assuming that your throughput is low and there is little confidence in demand growth for the product.

Certainly, the dynamics of market demand for a product would need to be well understood to risk a long-term agreement, say 3 to 5 years, with a contract warehouse operator. Moreover, if growth is expected and the user has warehouse expertise, private warehousing becomes a viable competing alternative.

Of significant importance will be to evaluate

- User's warehouse operating expertise
- User's expertise in managing personnel
- Current and future demand for product
- User's level of financial and other risk tolerance
- Need for control over product during distribution
- Fixed and variable costs
- Value-added and high-quality service requirements

## Case Study: Warehouse Selection and Distribution Quality

The following case study emphasizes the analysis aspects of selecting between warehouse alternatives. A series of analyses and solutions to the case follows the scenario.

The case pertains to the problems encountered by Sugar Creek Candy Company during the initial four months of new warehousing and distribution services utilizing Sweet Deal Distribution, Inc. (SDD), a public warehouse serving multiple and varying product clients.

The scenario helps to stimulate thinking about differences between warehouse alternatives. When comparing warehouse alternatives, differences must be evaluated both from a quantitative and qualitative perspective.

Service failure costs can mean the difference between a successful and profitable warehouse selection and a costly poor selection. Therefore, the user should obtain service quality performance records for each alternative warehouse under consideration and factor in the costs of even the slightest nonperformance percentage and cost. A supplier performance index illustrates the cost percentage increase due to service failures so that meaningful comparisons can be made in addition to comparing the quoted price per unit throughput.

Warehouse comparisons must consider the fixed and variable costs between the three general warehouse alternatives. Private warehousing has high start-up costs associated with fixed costs prior to the first unit of throughput. However, a savvy private warehouse operator should exercise greater control over service and variable costs. Contract warehouse costs may lie somewhere in between that of the public and private costs. Conducting pairwise comparisons between the fixed and variable cost for alternatives, a user may establish the level of product activity associated with the lowest cost provider. This is accomplished with a supplier point-of-indifference cost/unit analysis.

Space considerations are also evaluated in the situation and solution. Warehouse space needs should be evaluated in cubic feet and in square feet. The impact of aisle space, racks, and honeycombing on the overall designed space and space availability for product storage and movement must be considered.

## Sweet Deal Distribution, Inc. (SDD)—Warehouse Selection and Distribution Quality Case

From the Perspective of George Hatcher, V.P. of Distribution, Sugar Creek Candy Company

George Hatcher, the V.P. of Distribution for Sugar Creek Candy Company, just put down the phone and pondered the latest in a long list of growth and quality problems he has uncovered at Sugar Creek's newly hired public warehouse, Sweet Deal Distribution, Inc. (SDD). The call had been to schedule a meeting with the executives of SDD to discuss the options available to fix the problems. If the problems continue, Sugar Creek may be forced to give SDD a 30-day move-out notice. He knew now that he had to dig deeper for solutions to SDD's quality problems. He wanted to have his facts together before blaming SDD for all the service problems, so, he set out to analyze the situation from several aspects.

# SDD

Sweet Deal Distribution, Inc. (SDD), a 1.5 million sq. ft. public warehousing company, has been located in Atlanta, GA, for 20 years. SDD's first customer was a large consumable packaged goods manufacturer for whom SDD performed receiving, inventory storage, and shipping. Throughout the 1990s, SDD's client base expanded to more than 120, and the firm catered mostly to manufacturers of consumable packaged goods. Along with product storage, SDD also had an extensive freight consolidation program. This was especially attractive to clients because on their own no single client had enough volume shipping to a single consignee to warrant truckload service. SDD's capability to combine multiple small orders from multiple manufacturer clients and ship the orders together to common consignees allowed SDD to offer individual clients lower rates for each order. The consolidated rates were lower than what a single manufacturer could have negotiated with a less-than-truckload (LTL) carrier. Throughout the years, SDD gained a reputation in the Southeast United States as providing personable and high-quality service at the lowest competitive prices.

SDD receives shipments in the afternoons, picks orders at night, and ships orders in the mornings and some throughout the day. Product is received, checked, and put away in dedicated product zone locations. Orders are picked throughout the night one order at a time utilizing a manual paper system. Pick sheets and stock movement forms are entered into the system the following morning. Annual inventories cause a 3-day shutdown of the facility to count stock and reconcile inventory reports to the physical product count.

SDD's management structure is relatively flat and consists of the owners, Chief Operating Officer, vice president of Sales, finance manager, operations manager, customer service manager, traffic manager, management information systems manager, and an inventory control manager. Traditionally, SDD's strategic focus has been to provide low-cost warehousing and distribution services.

Order receipt and customer service primarily consists of receiving orders via client information system, phone, fax, courier, and sometimes drivers. SDD and several clients have established EDI capabilities for order transmittal and shipping verification. The traffic department handles order consolidation, routing, carrier negotiations, and claims processing. Warehouse operations are labor-intensive and consist of lift operators, checkers, stock locators, clerks, and supervisors. Radio frequency (RF) barcode technology was recently adopted for product receipt and storage. Product is shipped the day after receiving the customer's order.

SDD's reputation, years in the business, low-cost strategy, and a polished marketing pitch recently has led to three long-term contracts with other large candy and consumer goods manufacturers. The firm has quickly grown to approximately 1.5 million square feet located within four buildings. RF capabilities are available in two of the buildings.

## Situation

Problems began for SDD the day of contract agreements with the new customers. All three clients wanted to begin operations with SDD during the same month. SDD's owners were convinced, and convinced the clients, that this would be no problem. SDD created implementation teams to assist each new client.

The three clients operated under different strategic philosophies, had major customized product handling needs, and radically diverse communication and system requirements. Each required that SDD have the capability to totally integrate with the clients' processes, technologies, and people. SDD ensured that this was possible and gave the commitment to long-term partnerships with each of the consumable packaged goods product manufacturers. Exhibit A contains the information utilized by one of the new clients, Sugar Creek Candy Company, to make its decision to hire a public warehouse such as SDD for its distribution needs. The following are services that all the new clients require:

- Same day order receipt and shipping
- Same day delivery receipt verification
- Computer-to-computer communication
- Intensive cross-docking capabilities
- RF and barcode capabilities to track product throughout the warehouse
- Real-time inventory update, tracking, and reporting capabilities
- Daily cycle counts to replace annual inventories
- Quality program by which the DC measures all aspects of productivity and daily reports performance in each area

SDD began customizing to the customers' systems requirements and began hiring and training labor for handling the new business. The new clients began closing down multiple distribution and warehousing facilities and began operating throughout the Southeast with SDD as its single centralized distribution center. SDD's management ensured its current customers that the personable service, of which they were accustomed, would not be disrupted and that they would be treated equally important as the new larger clients.

SDD's customer service manager (CSM), Rich Lynder, became aware of possible major problems with handling the new business during yesterday's follow-up meeting at one of the new client's headquarters. The client was under the impression that all information systems were up and running, tailored process were in place, and workers were trained in the special handling needs. Moreover, someone in SDD's management informed the client that specialized storage bins and customized racking systems were installed in the warehouse. Rich Lynder knew better and informed the customer that the items were being worked on but were not yet in place. The client became alarmed that the systems were not in place and operational after 4 months of business.

In less than a year, SDD's plans began to fail. Product from the three new clients began filling the additional 300,000 sq. ft. of space. SDD's operations had virtually doubled overnight and orders began flowing into the DC for shipment. SDD's internal product tracking system proved inadequate as orders were lost in-house, shipment delays piled up, and inventory accuracy was approximately 82 percent. To make things worse, it seems that the estimated space requirements for the new business was inaccurately low and caused severe delays in putting away inbound product. Sugar Creek Candy Company (a new client manufacturing and marketing chocolate candy to wholesalers and retailers) estimated a total annual volume of 5.5 million cases (expected to increase next year to 9 million cases) at 12 inventory turns per year. The current maximum storage requirement at a given time period is equal to 1 month of cases. It was estimated by SDD that Sugar Creek would require 130,000 sq. ft. of temperature-controlled space to maintain the freshness of its product. Exhibit A contains a description of Sugar Creek's product to be stored.

Monster Merchandising, Inc. (a new client of dry goods) estimated a total annual volume of 2 million cases at 8 inventory turns per year. Happy Home Products, Corp. (a new client of small appliances) estimated a total annual volume of 40,000 pallets (containing four high cube cases each) at four inventory turns per year.

SDD began hiring additional laborers from a temporary service to keep costs low and manage the operation. This worked in the past when additional temporary labor capacity was required to meet the seasonal demand increase for a few clients. Only minimal additional management and supervisory staff were hired. Clients began requesting additional value-added service not specified in the contract. SDD made the decision to reduce the carrier base from 30 to 3 core carriers, and the primary core carrier was new to SDD's operation. The core carriers were not prepared to handle the special handling and checking requests of the new freight. Service levels began to diminish, and the core carriers began to ask for adjustments to the contract rates. Voice-mail and e-mail was added to SDD's communication network; however, customer service representatives began to use the system to avoid customers, and therefore, the problems exponentially multiplied. Previous consignee field visits to retailers, designed to keep a "finger on the pulse" of concerns of the ultimate customer, were abandoned. Prior to the new business, the previous customer service manager maintained bimonthly contact with retailers to correct any distribution problems and maintain a positive customer perception.

After 4 months of operations, things began to settle for SDD and the three new clients. Still, the warehouse space and systems were not meeting all needs of the new customers; cost-savings anticipated were not realized; and the primary core carrier had discontinued its contract obligations. To add to the stress, SDD was losing money on all the new business. The three new clients claimed that their warehousing costs were above the estimated cost of 5.83 cents per case received, 8.07 cents per case handling, and 5.10 cents recurring storage charge per case (based on average monthly inventory) because of SDD's poor quality service. For example, Sugar Creek Candy Company tracked customer returns, damaged and destroyed cases received at retailers, back ordering costs due to stock-outs, and late deliveries from SDD's facility. Exhibit B contains the "nonperformance" costs estimated by Sugar Creek.

SDD's owners were still convinced that the firm could handle the business and believed overall their customers were happy with the service provided. George Hatcher knew better. George felt that a completely new analysis was needed pertaining to warehouse space needs, product layout, nonperformance costs, and a breakeven analysis. George expects Rich Lynder of SDD to perform similar analyses to convince Sugar Creek to remain with SDD instead of switching to a contract warehouse or operating their own private warehouse. (See Exhibit C for alternative warehouse cost data.)

George hopes that the results of the analyses can help Sugar Creek and SDD in developing a service recovery plan, but it may result in Sugar Creek taking the business to Competition Contracting, Corp., a competing contract warehouse.

## The Immediate Challenge
George, the V.P. of Distribution for Sugar Creek Candy Corporation wanted to be well informed prior to meeting with SDD to discuss the candy company's recovery alternatives. He set forth to complete a thorough case analysis of the issues.

## Exhibit A: Sugar Creek Candy Company's Product Information
- There are 100 different product codes pertaining to Sugar Creek's business with SDI.

- The most popular items make up approximately 20 percent of the product codes and comprise approximately 80 percent of daily orders. Approximately 500 orders are shipped daily.

- The only product description available, however, is provided here.

  - Items are to be stored on pallets, in racks, five pallet positions high in each bay location.

  - Case size = 2 ft. width x 1.5 ft. length x 1.5 ft. height.

  - Palleted = 48-in. x 40-in x 3-tiers high.

  - The aisle allowance is estimated from past experience to be 20 percent.

  - The honeycombing allowance is estimated from past experience to be 12 percent.

  - The pallet height is 6 in.

- The clearance between pallets (side-by-side) is 4 in.

- Due to the racking system, the space between pallets (above each pallet) is 12 in.

- SDD estimated the need for 130,000 sq. ft. of temperature-controlled space for Sugar Creek's product.

## Exhibit B: Sugar Creek Candy Company's Estimated Cost of SDD Poor Service Quality

| Event | Cost/event | Number of Occurrences |
|---|---|---|
| Returns | $ 100 | 200 |
| Destroyed case of product | $ 150 | 298 |
| Damaged case salvaged at retailer | $ 95 | 175 |
| Back-order | $ 65 | 220 |
| Late delivery | $ 100 | 190 |

## Exhibit C: Sugar Creek Candy Company Information: Warehouse Alternative Costs

Private warehouse: Estimated Annual Cost for operating (following).

Total Annual Warehouse Cost = $1,885,800

Total variable cost = $385,000

35 percent cost for labor/supplies

32 percent cost for manager/section supervisors/office administrator/clerks

16 percent cost for storage maintenance

17 percent cost for machinery

Total fixed cost = $ 1,500,800

69 percent construction/lease facility cost

16 percent overhead

15 percent information management systems-WMS

Contract warehouse: Estimated Annual Cost for hiring Competition Contracting, Corp. a competing contract warehouse company (following).

Total Annual Warehouse Cost = $1,244,900

Total variable cost = $814,900

39 percent cost for labor/supplies

31 percent cost for manager/section supervisors/office administrator/clerks

10 percent cost for storage maintenance

20 percent cost for machinery

Total fixed cost = $ 430,000

60 percent construction/lease facility cost

10 percent overhead

10 percent information management systems—WMS

20 percent profit contribution

## Case Solution: Sweet Deal Distribution, Inc. (SDD)

Warehouse Selection and Distribution Quality Case

General answers:

1. From the point-of-indifference analysis, Sugar Creek Candy Company was correct in selecting a public warehouse such as SDD.

2. However, when accounting for SDD's service failures (nonperformance) costs, it would have been less costly had Sugar Creek Candy Company hired Competition Contracting, Corp. to handle its business, assuming Competition Contracting had significantly fewer service failures.

3. The warehouse capacity for Sugar Creek Candy Company's product was calculated without considering honeycombing. This is why the productivity is low and inventory accuracy is poor.

4. In hindsight, Sugar Creek Candy Company should have realized the capabilities of SDD would not have been enough to service the business in a high-quality manner. Sugar Creek Candy Company wanted dedicated service, equipment, and personnel to manage its distribution needs. It was not satisfied with sharing space and resources with other warehouse tenants. Issues pertaining to few skilled managers, not enough well-trained steady employees, poorly advanced processes, inadequate systems for the capacity requirements, and overall too high expectations have caused SDD's failure.

5. Sugar Creek Candy Company would be advised to move its business to Competition Contracting. Before this is done, Competition Contracting needs to create a plan for reducing warehousing cost while improving the service to its wholesale and retail customers.

## Analysis of Warehouse Alternatives Based on Fixed and Variable Costs (Establishing Total Cost Points-of-Indifference)

**Private:**  FC = $ 1,500,800

VC = $ 385,000 or ($ 0.07 VC per case @ 5.5 mil. cases)

TC = $ 1,885,800

Cost/unit = $ 0.34 or (TC / Annual Cases = $ 1,885,800 / 5.5mil. cases)

**Contract:** FC = $ 430,000 (contract guarantee)

VC = $ 814,900 of ($ 0.15 VC per case @ 5.5 mil. cases)

TC = $ 1,244,900

Cost/unit = $ 0.23 or (TC / Annual Cases = $ 1,244,900 / 5.5mil. cases)

**Public:**  FC = $ 000 (assumes portion included within per unit VC rate)

VC = $ 0.19

TC = $ 1,045,000 or [($ 0.00) + ($ 0.19) (5.5 mil. cases)]

Cost/unit = $ 0.19

$$\text{PRIVATE} = \text{CONTRACT}$$

$$(FC) + (VC) X = (FC) + (VC) X$$

$$(\$\,1{,}500{,}800) + (\$\,0.07)\,X = (\$\,430{,}000) + (\$\,0.15)\,X$$

$$(\$\,0.08)\,X = \$\,1{,}070{,}800$$

$$X = 13{,}385{,}000 \text{ cases or indifference point, which total cost equals (}\$\,2{,}437{,}750\text{).}$$

**Part of Answer: Use Contract up to 19,652,421 cases and Private thereafter.**

$$\text{PRIVATE} = \text{PUBLIC}$$

$$(FC) + (VC) X = (FC) + (VC) X$$

$$(\$\,1{,}500{,}800) + (\$\,0.07)\,X = (\$\,0.00) + (\$\,0.19)\,X$$

$$\$\,1{,}500{,}800 = (\$\,0.12)\,X$$

$$X = 12{,}506{,}667 \text{ cases or indifference point which total cost equals (}\$\,2{,}376{,}267\text{).}$$

**Part of Answer: Use Public up to 11,693,486 cases and Private thereafter.**

$$CONTRACT = PUBLIC$$

$$(FC) + (VC) X = (FC) + (VC) X$$

$$(\$ 430,000) + (\$ 0.15) X = (\$ 0.00) + (\$ 0.19) X$$

$$\$ 430,000 = (\$ 0.04.) X$$

$$X = 10,750,000 \text{ cases or indifference point, which total cost equals (\$ 2,042,500).}$$

**Part of Answer: Use Public up to 10,750,000 cases and Contract thereafter.**

**Final Answer: USE PUBLIC AT THE CURRENT 5.5 MILLION CASES (TC = $ 2,042,500)**

## Analysis of Nonperformance Costs Using an SPI (Supplier Performance Index) Analysis to Adjust for Service Failures (Establishing Total Costs That Include Adjustments for Poor Quality Service)

Sugar Creek Candy Company's Estimated Cost of Poor Service Quality from SDD Warehousing

| Event | Cost/event | Occurrences | Extended Cost |
|---|---|---|---|
| Returns | $ 100 | 200 | $ 20,000 |
| Destroyed case of product | $ 150 | 298 | $ 44,700 |
| Damaged case salvaged at retailer | $ 95 | 175 | $ 16,625 |
| Back-order | $ 65 | 220 | $ 14,300 |
| Late delivery | $ 100 | 190 | $ 19,000 |
| TOTAL Non-Performance Costs | | | $114,625 |

**Part of Answer: Total dollars spent with SDD this 4-month period:**

(5.5 million cases) / (12 months) = 458,333 cases per month

4 months = 1,833,333 cases shipped. Cost per case through public warehouse = $0.19/case

Total Dollars Spent this four-month period:

(1,833,333 cases) ($0.19) = **$348,333**

Adjusted Total Dollars Spent:

(Dollars spent with SDD this four-month period, see previous calculation) + (Nonperformance costs):

= ($ 348,333) + ($114,625)

= $ 462,958

Supplier Performance Index associated with SDD's charge per case plus non-performance costs:

= (Adjusted Total Dollars Spent) / (Total Dollars Spent with SDI)

= ($ 462,958) / ($ 348,333)

= 1.33 SPI (Supplier Performance Index)

Final Answer: An SPI of 1.33 means that Sugar Creek Candy Company's per unit cost from SDD is actually 133 percent of the price per unit that SDD is charging Sugar Creek. That is, due to the nonperformance costs, it is actually costing Sugar Creek 33 percent more per case using SDD as its warehouse.

Cost per case through public warehouse = $0.19/case

SPI = 1.33

Actual cost to do business with SDD: ($ 0.19) (1.33) = $ 0.25/case

$$\text{CONTRACT} = \text{PUBLIC}$$
$$(\$430,000) + (\$0.15)\ X = (\$0.25)\ X$$
$$\$430,000 = (\$0.10)\ X$$
$$X = 4,300,000 \text{ cases or indifference point, so total cost equals } (\$1,075,000).$$

Factoring in the nonperformance costs, and realizing it needed dedicated space, dedicated handling resources and personnel, and customized processes, Sugar Creek Candy would be advised to use the competing Contract warehouse services assuming it had significantly less service failures. Up to 4,300,000 cases the total cost for the public warehouse option would be less than for the contract warehouse choice. For example, at 4,000,500 cases the costs would look like this:

Public = ($0.25) (4,000,500 cases) = $1,000,125

Contract = ($430,000) + ($0.15) (4,000,500 cases) = $1,030,075

However, at 4,340,000 cases the total cost for the contract warehouse would be $4,000 less than for the public warehouse choice during the 4-month period.

Public = ($0.25) (4,340,000 cases) = $1,085,000

Contract = ($430,000) + ($0.15) (4,340,000 cases) = $1,081,000

The service quality of the contract warehouse would have to be estimated so that an appropriate SPI could be applied for the contract warehouse, too. The data in the case does not allow for this calculation, but the manager could ask for proof of service performance based on the contract warehouse's previous clients served.

**Final Answer: Use Contract Warehousing Considering the Nonperformance Costs at the Current 5.5 Million Annual Cases (TC = $1255,000) Because Due to the Poor Service Quality, SDD Is Actually Costing Sugar Creek Candy $120,000 More Annually Than the Contract Alternative.**

## Analysis of Sugar Creek Candy Company's Space Requirements

The example was created based on Tompkins and Smith (1998, 2013, p.245)

Provided from text: 5.5 million cases / 12 months = 458,333 cases per month and the maximum cases on hand expected at any given time period..

**Exhibit B**

**Sugar Creek Candy Company's Product Description**

- There are 100 different product codes pertaining to Sugar Creek's business with SDI.
- The most popular items make up approximately 20 percent of the product codes.
- The only product description available, however, is provided here.
- Items are to be stored on pallets, in racks, five pallet positions high in each bay location.
- Case size = 2 ft. x 1.5 ft. x 1.5 ft. (height) = 4.5 ft$^3$
- Palleted = 48-in. x 40-in x 3-tiers high = 60.12 ft$^3$ (12 cases/pallet)

**Part of Answer:**

**(Note: palletized unit dimensions in feet = 4 ft. x 3.34 ft. x 4.5 ft.)**

**(Note: full pallet height here = (3-tiers high) (1.5 ft. height per case) or 4.5 ft.**

The aisle allowance is estimated from past experience to be 20 percent.

The honeycombing allowance is estimated from past experience to be 12 percent.

The wooden pallet height is 6 in.

The clearance between pallets (side-by-side) is 4 in.

The racking system requires 12 in. of space above each pallet.

**Part of Answer:**

(Pallet Width + Clearance between Pallets) (Pallet Length)

X

{[(Case Height) (Number of Tiers High of Cases per Pallet)] + Wooden Pallet Height + Space Above Each Pallet for Racking System}{Number of Pallet Positions High}

= Cubic Feet Needs for a Complete Bay of Product

**Part of Answer:**

**Sugar Creek's space needs for one bay of product:**

= (4 ft. + 0.33 ft) (3.34 ft.) {[(1.5ft.) (3)] + 0.5 ft. + 1 ft.} {5}

= (4.33 ft.) (3.34 ft.) (6 ft.) (5)

= **521 ft³ for a complete bay of product** (Or 5 pallets, or 60 cases, that is 12 cases per pallet times 5 pallets high)

**Part of Answer:**

With the inclusion of aisle and honeycombing:

= (Space Needs per Bay) / [(1-Asile Space %) (1-Honeycombing Space %) (Cases per Bay)]

= (521 ft³) / [(1- 20 %) (1- 12 %) (60)]

= (521 ft³) / [(.80) (.88) (60)]

= (521 ft³) / (42.24)

= 12.33 ft³ per case

**Part of Answer:**

Therefore, the total temperature-controlled storage space required for Sugar Creek Candy Company's product at the maximum number (458,333, see above) of cases expected at any one time is

= (458,333 cases) (12.33 ft³ per case)

= **5,651,246 ft³ needs having a 30 ft. total bay height including pallet and rack.** That is, {(0.5' wooden pallet height) + [(3) (1.5) palletized cases height] + (1' rack clearance above each pallet)} (5 pallets high) or a total of 30 ft. bay height.

**Part of Answer:**

Square ft. needs are computed as follows:

- Maximum cases expected on hand at a give time:

    = 458,333 cases (see previous calculations).

- Maximum pallets expected on hand at a give time:

    = (458,333 cases) / (12 cases per pallet)

    = 38,194 pallets

- Maximum pallets on the floor of the warehouse expected at any given time:

    = (38,194 pallets) / (5 pallets high per bay)

    = 7,639 pallets on floor

- Square ft. needs per pallet:

    = (pallet width + clearance between pallets) (pallet length)

    = (4 ft. + 0.33 ft.) (3.34 ft.)

    = 14.46 ft. needed per pallet footprint.

- Square ft. needs per pallet when allowing for aisle space and honeycombing:

    If aisle space allowance = 20 percent

    If honeycombing allowance = 12 percent

    If square ft. needs for one pallet footprint = 14.46 ft.

    Then,  Aisle space     = (20 %) (14.46 ft.) = 2.89 ft.

    Honeycombing = (12 %) (14.46 ft.) = 1.74 ft.

    Therefore,

    (Square Feet Space for One Pallet Footprint) + (Aisle Space) + (Honeycombing)

    = (14.46 ft.) + (2.89 ft.) + (1.74 ft.)

    = 19.09 ft. space needs per pallet considering aisles and honeycombing

**Final Answer:**

(7,639 Pallets on Floor) (19.09 Feet per Pallet Considering Aisle and Honeycombing) = 145,829 Square Feet Needs for Maximum Number of Pallets at Any Given Time of Sugar Creek Candy Company's Product

SDD allowed and quoted only for pallet and aisle space but forgot to consider honeycombing space when estimating the space needs at 130,000 square ft. It is off by 15,829 square ft.

In conclusion, the lack of honey combing space caused significant bottlenecks when moving product into and out of storage. Other consequences of improper space allocation include more travel time to putaway product because the operator must search for the one open space. Or the WMS space assigns the next closest space, but it may be well away from the receiving dock. Blocked SKUs by another item often occurs when space is critically limited. Damaged product, missing product, and employee safety are all at risk. Although honeycombing may seem like "unused space," a percentage of buffer space helps to relieve the issues discussed.

## Key Terms

- Contract Warehouse
- Cost Point of Indifference
- Distribution Center
- Fixed Costs
- International Standards Organization (ISO)
- Kitting
- Mixed Pallet
- National Association of Government Contractors (NAGC)
- Postponement
- Private Warehouse
- Public Warehouse
- Shrink-wrap Tunnel
- SKU
- Special Handling
- Start-up Costs
- Storage Agreements
- Third-Party Logistics Provider (3PL)
- Throughput

- Value-Added Services (VAS)
- Variable Costs
- Warehousing Education and Research Council (WERC)
- Warehousing Strategy
- Lean Six-sigma
- Council of Supply Chain Management Professionals (CSCMP)
- Cross-docking
- Merge-in-transit
- Freight Consolidation
- Deconsolidation
- Freight Payment
- Labeling
- Foreign Trade Zone (FTZ)
- Importer
- Customs Duties
- High-cube
- Relabeling
- Point of Indifference
- Supplier Performer Index
- Honeycombing

## End Notes and Suggested Readings

Akerman, K. B. (1997, 2012), *Practical Handbook of Warehousing*, 4th ed., Chapters 2, 3, and 10, Chapman and Hall, New York, NY.

Coyle, J. J., C. J. Langley, Jr., R. A. Novack, and B. J. Gibson. (2013) *Supply Chain Management: A Logistics Perspective*, 9th ed., Chapter 11, South-Western, Mason, OH.

Kohn, J. W., M. A. McGinnis, and J. E. Spillan, (2009), "A Longitudinal Study of Private Warehouse Investment Strategies," *Journal of Transportation Management*, Vol. 21, No. 2: pp. 71–86.

Murphy, P.R. and D. F. Wood. (2011), *Contemporary Logistics*, Chapter 10, Prentice Hall, Upper Saddle River, NJ.

Napolitano, M. and Staff at Gross and Associates (2003), *The Time, Space and Cost Guide to Better Warehousing Design*, 2nd ed., Chapter 3, Distribution Group, New York, NY.

Tompkins, J. A. and J. D. Smith, (1998, 2013) *The Warehouse Management Handbook*, 2nd ed., Chapter 9, Tompkins Press, Raleigh, NC.

# 4

# DESIGN AND LAYOUT

## Introduction

This chapter discusses how product factors, transaction throughput, and available supporting equipment and technology influence and shape warehouse design and layout. Storage areas, dock space, and equipment available combine with product factors and customer requirements to influence warehouse design, layout, and management. Envisioned future business plans and expansions also influence this area of discussion.

## Design and Layout

Students enrolled in a warehousing class were faced with a challenging case study. A regional food bank was considering constructing a 30,000 square foot building to accompany the existing 30,000 square foot facility that was operating at maximum capacity. The students were asked to evaluate the layout of the current warehouse and make recommendations to improve product flow and storage. Students were also asked to provide insights into the need for the second facility. Research disclosed several underlying problems with product flow, storage, and process that kept the existing warehouse from efficiently utilizing the maximum space available.

The building contained two receiving doors in the center front of the building and two shipping doors on the right side of the building. Product flowed through the facility in a dogleg or L-shaped pattern. Donated product primarily was received from retail stores or distribution centers (DC) in mixed and full pallet quantities. One-third of the cases were opened and individual items were shelved for shoppers to select. Other orders would entail shipping in full pallets, mixed pallets, or individual cases.

Of the many issues, several surfaced that if corrected would significantly expand the capacity availability inside the warehouse. The added efficiency and space availability would eliminate the need for building the planned second facility.

There appeared to be little consideration for categorizing and managing product groups according to case or pallet size, volume, or velocity. Space was allocated for shelving products in a manner similar to that of a grocery store. It appeared to be large considering the light flow of customers through the facility and having calculating the percentage of honeycombing (the ratio of empty to total available shelf product facings or locations). Basically, there was no clear plan or process for positioning product in the warehouse or efficiently moving the product through the order fulfillment process. Product was placed in aisles blocking other product and available rack space and prohibiting the flow of product and equipment moving through the facility.

## Importance of Space and Time Within the Warehouse

Planners of product flow through the warehouse should be obsessed with evaluating the utility of the available space and reducing the time necessary for moving product through the facility. More important, space is one primary factor when calculating cost and price. You must understand the product and customer characteristics that influence the efficient utilization of space and time expended to receive, putaway, pick, and load products. Weight, cube, packaging, variety, velocity, and other factors related to product handling are critical factors.

### Weight, Cube, Packaging, and Other Product Factors

Product characteristics such as weight and cubic dimensions often have an influence on the stacking height and physical location of products within a facility. Pallet quantities of cereal may be stacked higher utilizing slip-sheets compared to stacking pallets of quart-sized glass bottles of cooking oil. Boxes of cereal may not require a racking system, and therefore, may be stacked to the maximum height of the facility and efficiently utilize the maximum square footage of the floor. However, the weight and packaging of some products may dictate the utilization of a rigid racking system to handle the weight of the product and pallet when desiring high-cube storage utilization. Racks reduce the capacity available for storing products, yet the entire designed capacity of the facility must be factored into the cost of product storage. Therefore, racks may be necessary to improve cube utilization by allowing product to occupy vertical pallet positions.

Several major retail distribution centers segment products into nonconveyable, nonstackable, and extremely bulky products. Such products may be located nearest the shipping doors so that the special handling required may be minimized when pressured to meet shipping schedules. In one extreme case the product required two people and a forklift to pick up and guide the product to the staging area.

Product characteristics are important when making decisions concerning product location within the warehouse.

# Variety

Single-item inventory would be most efficient because a simple first-in first-out (FIFO) process could be employed. Product case size and SKU velocity would not vary; consequently variety wouldn't be an issue. Product could be put away nearest receiving doors, and the handling process could replenish forward-picking slots with the newest product nearest the shipping doors.

However, warehouses managing inventory variety would need to evaluate the physical characteristics and ordering dynamics of each SKU. Doing so would allow managers to group similar products together so that each is managed in the most-efficient manner. For example, a product ordered once per month should not be positioned beside a product ordered daily. Inventory managers would utilize an ABC analysis; whereby, SKUs are grouped and positioned in the warehouse by the similarity of the products. Similarities may include order frequency, ease or difficulty in handling, cubic or density characteristics, or simply by items that often are ordered together.

Greater SKU variety requires greater space availability. It is inefficient to block one SKU with another SKU occupying the same bay location. Picking would take extra time to move the outside product to get to the inside product. Each SKU requires an individual bay or slot location. The greater the number of SKUs, the greater the number of bay and slot locations required.

# Velocity

Velocity pertains to the speed with which a product moves through the facility; from the time it is received until it is shipped. Fast-moving items would be placed together so that order fillers may have quick and easy access to those items ordered most often. SKUs ordered most often and in the greatest quantities would be classified as "A" items. Slower moving items would be grouped together as "B" items, and "C" items would constitute the slowest movers and should be positioned further away from the shipping area than "A" items. This way, pickers would have reduced travel time associated with the fastest moving items and could pick, stage, and load most efficiently those items moving most often. Locations within the warehouse that are less accessible or require more travel time to reach would contain items that are ordered less frequently and have lower velocities (B and C items).

Cross-docking would be an extreme case of high-velocity product movement. Increased velocity facilitates greater inventory turnover and consequently a lower cost per unit handled. For this reason and when possible, managers may decide to cross-dock freight. Items having little to no threat of obsolescence or those having lengthy shelf lives may be good candidates for cross-docking. This minimizes the handling and travel distance/time when filling orders. As managers expand cross-dock operations, more space will be allocated to the dock, whereas less will be allocated to storing inventory.

# Handling

Other product handling characteristics used to determine storage location within a warehouse include the equipment available and the capability to convey product through the picking process. Break-pack and other processes utilized to reduce the size of units ordered may require electronic picking tunnels, carrousels, A-frames, and other machinery to assist in the picking of small items rather than by case quantity. Figure 4-1 illustrates the sectioning of a warehouse based on the various product ordering and handling characteristics.

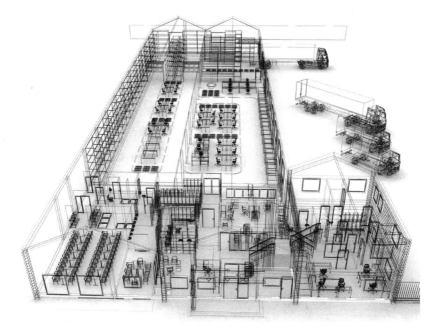

Figure 4-1   Warehouse design and layout (image courtesy of gwycech/shutterstock)

## Staging and Loading Docks

Loading docks are important areas to expedite receiving and shipping and have changed throughout the years. As previously indicated, cross-dock operations require generous dock space for managing the product sortation from inbounds and moving freight to orders staged for outbound shipping.

In some cases it is possible to pick a product and immediately load the product onto the trailer. This is especially possible when companies utilize a drop trailer program, whereby, an empty trailer is positioned at an outbound shipping door and loaded at the discretion of the warehouse manager. Live loads, or shipments where a driver waits for the shipment to be loaded, also are timely when product is picked and directly moved onto the awaiting trailer.

Other cases require outbound orders to be picked and staged nearest the outbound shipping door awaiting a scheduled pickup. Dock space may be configured to allow for the entire trailer loads to be staged according to the order loading sequence. Again, the ability to reduce loading time also reduces total supply chain lead time and cost.

Inbound dock space may be required to facilitate counting and inspecting inbound loads. One distribution center in Michigan requires that a new vendor's inbound products pass inspection on three consecutive loads received; after which the DC forgoes product inspection during receipt. An area within the receiving dock space is allocated for inspection of products and may include opening cases to inspect garments or placing small appliances on a machine that tests the durability of a product if dropped or shaken within specified ranges.

## Dedicated and Random Storage

Universities, especially the largest, have relatively large warehouse operations to manage the materials and goods required for supporting on-campus restaurants, housing, maintenance, administration, and faculty. One large university purchased so much copy paper that it received discounts from suppliers beyond that available to their State's government operations, so the State purchased paper from the university. The inventory of paper was so large that it demanded a designated location in the warehouse that would not change. Large orders for paper were received on a continuous basis, and warehouse personnel knew exactly where it would be placed in the warehouse.

Regular and large inventories of single SKUs are prime candidates for dedicating pallet slots and bay locations. This keeps putaway and picking personnel from traveling to multiple locations to receive or ship large case quantities of the item.

Dedicated storage plans help to ensure that personnel recall product locations during putaway and picking. This may help to keep inventory integrity and reduce process variations in operator travel time within the warehouse.

Randomized storage allows lift operators to place inbound products in the first available space nearest the receiving location. Receiving time is reduced, and with today's warehouse management systems, the items are easily located for shipping when required. The parent company of several global brands of apparel received SKU varieties in the hundreds throughout any given week. Rarely were inbounds full pallet quantities of single SKUs, and outbound orders consisted of pieces or "eaches" rather than cases or pallets of items. Random storage of all SKUs allowed receiving to place items in the first available slot location. Shipping was not impaired because it, too, had many SKUs on each order requiring them to travel throughout the warehouse to fulfill orders.

Dedicated storage is best-suited for large quantities of single SKUs, classes of products with similar handling and ordering characteristics, such as items requiring special handling because of their bulky and nonconveyable nature and materials that are hazardous.

The manager of the large university distribution center believed it would be advantageous to move from a dedicated storage of all SKUs to a randomized system because it adopted a new warehouse management system. After 3 months the manager realized that his employees had overridden the warehouse random storage assignments and placed SKUs back into the previous dedicated storage locations. The manager acquiesced, allowing the dedicated storage layout because he felt it was not a major financial concern to the university.

## Aisles

Aisles influence the efficient handling of product. Narrow aisles enable greater storage. However, the aisles within a distribution center of a major confectionary brand were uncustomarily wide. When asked about the width, the manager said it decided that it would pay off to have wide aisles to allow forklift drivers to accelerate to the maximum designed safe speed of each forklift. Putaway and picking times would be reduced and consequently total lead-time for the order would be reduced.

## Value Added Services (VAS)

As previously indicated, to compete, warehouses must provide services beyond traditional storage. Some create value in ways that were previously or traditionally performed by manufacturers. By performing value added services closer to the end customer, warehouses can provide products tailored to customer needs without delays associated with core manufacturing processes.

Following are warehouse value-added services that affect layout and design:

- Special handling
- Sorting
- Kitting
- Assembly
- Label generation
- Import/exporter customs services
- Repackaging
- Final configuration
- Light manufacturing

- Shelf-life management

- Vendor compliance management

- Ongoing maintenance and upkeep if required

The wiring harness example in Chapter 1 is a prime example of a warehouse operation altered to provide more specialized and contract value-added services to the manufacturer. Other value-added services may entail, for example, changing brand labels on garments that failed to sell at one retailer and were brought back to the warehouse to be relabeled and redistributed to another retailer. Affixing sales tags to garments and packaging products according to customer requirements are two value-added services that require additional warehouse space beyond the receiving and loading docks and storage.

## OS/D and Returns

Overages, shortages, and damages (OS/D) pertain to discrepancies. Product that requires cooperage because of damage to packaging in shipping or within the warehouse must be managed. The OS/D area is an area within the warehouse designated for recovering damaged product, holding product awaiting disposition, and product that was received in excess.

An order fulfillment center supporting a televised shopping marketer had a section of the warehouse dedicated for returns. Product was received from customers and required inspection, and the disposition of the product was determined. Product was assigned to OS/D for repackaging, sold to a salvage company, or destroyed.

## Customer Requirements

Space requirements often exist due to reasons beyond the physical product characteristics and critical value-added services. For example, in one temperature-controlled warehouse, a candy manufacturer paid for space that was not shared by other tenants. When product on-hand dipped below the 30,000 square feet space availability, the client required that the space remain empty until filled with the company's candy. Public warehouse prices are based on the flexibility to assign space to various clients as demands change for each customer. Certainly, the dedicated space for the candy product would influence the productive space available for other clients.

## Equipment Systems for Movement and Storage

Similar to affixed racking systems, mechanical product movement such as conveyors or other handling equipment require warehouse space. One trade-off exists between space and order-fill time or lead-time within the warehouse. For example, a carousel may allow small items to be brought directly to an order picker. The system reduces the travel time for the picker as the product is brought to the picker. Although many SKUs are stored within the carousel, it does not allow for bulk SKU storage. Pick tunnels that utilize pick-to-light systems and processes also require machinery space. And while the operator moves to the pick location, the system is especially good for picking pieces/eaches from break-pack SKUs. Products are picked in a sequence from closest to furthest from the picker. In this manner the picker moves in a straight line selecting product.

Automated Storage and Retrieval Systems (AS/RS), like the one pictured in Figure 4-2, enable the efficient employment of high-bay storage supported by narrow aisles and automated mechanical systems to store and move product. AS/RS systems are especially appropriate for full pallet quantities of single SKUs that are received and ordered in full pallet quantities. In addition, the automotive industry utilizes AS/RS systems to store and move automobile parts, such as quarter panels, hoods, trunks doors, and heavy and bulky parts.

Figure 4-2    Automated Storage and Retrieval Systems (AS/RS) (Image courtesy of Baloncici/shutterstock)

## Common Product Layout and Flows

With so many product and service variations, multiple warehouse layouts and plan possibilities exist. Some entail reserve and active picking areas. Reserve locations are designated for bulk product that sit at rest until warehouse personnel replenish forward picking areas, also called active picking locations. The process demands minimal personnel only in the majority of the stock for putaway and replenishment. Much of the activity in a warehouse occurs within the picking process. The reserve/active layout keeps personnel away from the majority of the product and contributes to employee safety and inventory integrity.

Bulk pick lines constitute one common process utilized in a public warehouse that distributes consumable products. A form of active/reserve, a bulk pick line is created during nonshipping hours. Bulk pick sheets are generated to guide pickers to pull at one time all cases of single items designated to ship on orders. In this manner, SKUs common across items are picked in bulk and staged in an active pick line. The next shift fills orders and loads trailers directly from the bulk pick line. The process reduces travel time associated with the picking and loading processes. Replenishment personnel are assigned to designated reserve locations, which helps to maintain familiarity and inventory integrity.

Zone picking is also common and limits employee exposure to unfamiliar inventory. In a distribution center supporting home shopping, product is situated within zones, and specific employees are assigned to each zone. Single order picking is conducted by employees having familiarization with the products in that zone. Products are picked by order and staged for outbound shipping. Each zone stages products to complete orders. Variations of standard practices are adopted and altered to customize the receiving and picking sequence for customers.

## Future Plans and Expectations

Warehouse layout and design also is influenced by the nature of future plans of the operator. Operators expecting increased throughput may plan for added space and potentially added doors to receive and ship. As activity increases it is expected that additional space will be necessary for added OS/D, storage, and dock space. Of course, operators must evaluate the extent of added value required for the increased inventory.

# Summary of Key Points

Layout and design of a warehouse facility is influenced by the dynamics of the product characteristics, ordering characteristics, machinery employed, client demands, expectations of future activity. It is extremely important for a warehouse operator to evaluate the following factors associated with managing the warehouse:

- Product throughput
- Product variety
- Materials handling equipment
- Receiving requirements
- Storage and cross-docking requirements
- Picking and staging requirements
- Value-added services and OS/D
- Future expectations of volume and handling needs

Several programs exist to perform what-if scenarios to evaluate different warehouse layout plans. Historic data is exceptional for determining the optimal or preferred layout of product and warehouse design.

## Key Terms

- A-frames
- ABC Analysis
- Active Picking Area
- Automated Storage and Retrieval Systems (AS/RS)
- Break-pack
- Bulk Pick Line
- Cross Docking
- Distribution Center (DC)
- Drop Trailer Process
- Electronic Picking Tunnels
- First-In First-Out (FIFO)
- Forward Picking Area
- Honeycombing
- Inventory Integrity

- Inventory Turn Over
- Kitting
- Live Loads
- Mixed Pallet
- Obsolescence
- Pick Tunnel
- Pick-to-light System
- Picking
- Product Flow
- Putaway
- Reserve Picking Area
- SKU
- Sortation
- Sorting
- Value-Added Services (VAS)
- Inventory Velocity
- Zone Picking

## Suggested Readings

Akerman, K. B. (1997, 2012), *Practical Handbook of Warehousing*, 4th ed., Chapter 15, Chapman and Hall, New York, NY.

Baker, P. and M. Cannessa, (2009), "Warehouse Design: A Structured Approach," Production, Manufacturing and Logistics, Vol. 193, No. 2: 425–436.

Tompkins, J. A. and J. D. Smith, (1998, 2013) *The Warehouse Management Handbook*, 2nd ed., Chapter 9, Tompkins Press, Raleigh, NC.

# 5

# PERSONNEL

## Introduction

Hiring, training, and retaining good people are salient factors in the success of warehouse operations. The labor force skill level and motivation level defines the upper limits of efficiency that is attainable. Competent leadership is imperative in today's business environment. Warehouse managers and supervisors must not only be technically grounded in all aspects of warehouse operations, but they must also be skilled personnel managers and motivators. A trained workforce requires resources and a dedication of management to continually enhancing employee skill sets. Together managers and the workforce establish the culture of the respective warehouse. A well-structured and well-understood performance appraisal system can assist in reinforcing good performance, help with continual-improvement objectives and help correct deficient behavior.

## Personnel

It could be said that a warehouse is only as good as its personnel. Many times warehouse employees provide the final-touch point of products and orders prior to customer receipt. All the product design and marketing invested in the success of the product may be void by a warehouse that performs poorly during final distribution. Personnel in managing the materials and work-in-process warehousing and distribution also impact the next-in-line customer (internal customer or external customer to the organization).

Supply chains are advised to embrace the role of warehouses in the quality and cost of total supply chain management. Research indicates that measures must be taken to ensure the quality of personnel and processes they perform to achieve successful goals of the warehouse. This chapter provides direction to managers wanting to hire, train, and harvest the most from their frontline warehouse employees at competitive costs and highest safety.

# Labor

Historically, it was necessary for some manufacturing lines to increase production capacity by temporarily moving warehouse workers to the production line when assembly demand required. Today's production lines require more technical skill to manage the processes and technology of production, so warehouse personnel must look to provide value in storing and handling materials and finished goods. Progressive warehouses of today must locate, hire, and provide the training and support to produce highly skilled employees to manage process integrity and to interact with technology in ways unlike in the past. This requires a change in perspective from traditional warehouse management thinking.

A Midwest contract warehouse supporting large-scale retailing requires each frontline employee to undergo intensive training, job shadowing, and testing prior to being assigned to load trailers. This particular company floor loads trailers to better utilize all possible cube and to reduce the overall cost per employee and trailer. Productivity is of great importance and is determined by the speed of loading and capacity utilization of a trailer. One value-add activity performed by the contractor included placing radio frequency identification (RFID) tags on the products so that the end retailer could better track and trace the inventory when in its control.

Warehouse employees must perform duties under time pressures and with high levels of accuracy. Competitive warehouse operators have adopted technology to assist frontline workers in selecting, checking, lifting, and moving product. Employees must achieve a level of competence when interacting with technology and ensuring that standardized processes retain their integrity. It is no longer enough to have the brawn to lift and move heavy boxes.

Operations labor includes product handling personnel. This category manages the physical movement of product from the time of unloading to ultimate loading of outbound orders. Receiving and shipping personnel must be skilled in operating various specialized and standard forklifts or walkies (pallet jacks) when handling pallet quantities of products. Lift operators interact with information technology to store or pick product according to the appropriate inventory position in the warehouse. Task interleaving is conducted by warehouses having warehouse management software sophisticated enough to plan, integrate, and assign various warehouse tasks to a pool of materials handlers taking into consideration the location of each operator within the warehouse.

A new warehouse facility began utilizing task interleaving; whereby, a lift operator would unload a pallet of product and store the product in the computer-assigned slot location. After the lift operator informed the warehouse management system (WMS) that the task was completed, the system would then assign the operator to a new task. Tasks were assigned based on the nearest operator to the next critical task. Therefore, the operator may be assigned to putaway a pallet but then be assigned to move a pallet of distressed product that, for example, is out of code date, from the reserve area to OS/D. From there, the system would assign the operator a new task.

Task interleaving is designed to reduce the cost of moving product throughout the warehouse by reducing the travel time of warehouse operations personnel. However, one high volume warehouse found that the total time for receiving a trailer actually increased as lift operators were assigned various tasks of which some included receiving the occasional pallet off of an inbound load. When asked, the workers reported that they had no conception of how well they were performing their duties; because they used to compare the number of trailers unloaded by each operator. This was no longer a means to compare performance between employees, and therefore, the healthy competition among workers was no longer a performance motivator. Management decided to assign unloaders only the task of unloading trailer loads to regain the competitive atmosphere. Immediately the trailer unloading performance improved.

Other warehouse operating positions outside of management include clerks and inventory control clerical-type personnel. A receiving clerk often interacts with truck drivers and with unloaders. Receiving clerks are responsible for inspecting and counting all inbound freight. They work with drivers to resolve discrepancies and when necessary provide instructions to drivers. Proof-of-delivery receipts are signed by receiving personnel, and any issues pertaining to discrepancies in count or condition of product are notated on the delivery documentation including on the bill of lading.

Outbound shipping clerks verify the count and condition of all outbound freight. They work with picking operators, truck drivers, and inventory control personnel when necessary. Outbound clerks must receive a truck driver signature on the bill of lading to indicate that the driver/carrier has received the freight in proper condition and in the designated quantities. Any exceptions that deviate from the original order are documented on the bill of lading by the shipping clerk and verified by the driver.

Progressive warehouses utilize electronic documentation provided via WMSs. The risks and exposure associated with losing paperwork are nearly eliminated. Drivers may also have electronic documents stored and transmitted through their on-board computers (OBC). Instructions and electronic documents may be communicated via OBCs between drivers and fleet managers.

Inventory control clerks participate in activities and processes designed to enhance inventory integrity. Inventory management systems have significantly improved the management of inventory. Inventory control clerks may perform scheduled physical inventories and reconcile discrepancies between the actual product in the warehouse and what the system indicates is on-hand. Discrepancies for one warehouse were significantly reduced as clerks began a daily process of physically counting the highest volume SKUs having shipped each day. Prior to instituting the process, called cycle counting, the SKUs had the greatest activity and most errors during shipping. Cycle counting allowed inventory personnel to identify miss-shipped product and reconcile the physical inventory on a daily basis. Responsibility for errors immediately could be assigned, and pickers could be counseled on how to improve their accuracy before the next day's picking cycle began.

OS/D employees manage product requiring special attention. Although termed overages, shortages, and damages (OS/D), some warehouses simply call an area for managing damaged product the rework area. Product received in excess of the expected amount will require a decision on disposition. Clerks must decide if the product will be received, rejected, or returned to the shipper. OS/D personnel work with inventory clerks to determine the future of the product. Rework may also entail inspection of product that appears damaged to determine the extent of damage and the potential for repackaging the product. Damaged product will be removed from available inventory and moved to the OS/D or rework area in the warehouse. Electronic or physical stock movement forms are generated to place the product on hold and in rework. After product is repackaged it will then be moved back into available inventory status and physically placed back into reserve or active inventory. It is the OS/D clerk's responsibility to manage the rework process and communicate with inventory control on the status of product in OS/D.

Professor Steve LeMay and colleagues provide a comprehensive and detailed depiction of the job descriptions and skills required of each warehouse position in *The Growth and Development of Logistics Personnel* (LeMay 1999).

## Supervision

Warehouse managers may supervise a section of the warehouse or an entire facility. Supervision of all aspects of the operation is necessary; from marketing to the warehouse floor. Interacting with frontline employees requires a combination of managing processes and people in the most effective ways to retain the best employees and formulate the most-efficient and effective processes. Following is a list of some core competencies required of warehouse managers:

- Decision making and delegating
- Planning and innovation
- Managing risk, safety, and security
- Communication
- Mentoring and coaching
- Counseling and resolving issues
- Personnel training and development

In addition, supervisors and managers must have the following skills:

- Supervisors must be skilled in managing and motivating groups of employees that are not always in sight during daily operations. They must understand each process within their responsibilities and help employees reach their highest level of accuracy and productivity possible. Doing so requires supervisors with strong skills in organization, communication, and time management.

- Managers will be responsible for budget creation and meeting expectations of the approved budget. Personnel assignment and equipment utilization can greatly impact the ability to meet budget. Managers must evaluate the scheduled daily inbound and outbound activity and any anticipated internal product movement or maintenance. Capacity cushions should be estimated to ensure that enough personnel are on hand to perform expected and unanticipated activities.

- Managers must be skilled in training, coaching, and counseling hourly employees. Progressive warehouses employ formal trainers for teaching new hires. Skilled trainers can then be called upon to assist supervisors in helping employees that require retraining or additional training. Coaching is often used by skilled supervisors to bring subordinates into alignment with expectations in productivity and professional behavior. Counseling is provided to employees that require critical direction to alter behavior patterns and as a final step to avoid dismissing an employee.

## Keys to Creating a Customer-Focused Warehouse Workforce (Section Derived from Keller 2007 and Keller and Ozment 2010)

Market orientation pertains to the clarity of focus that a business has with respect to the customer. Market-oriented firms understand customer wants and needs and center their business goals and propositions around the customer. Coupled with the firm's capability and willingness to tailor services and products to meet specific customer needs and expectations, a firm is said to be market-oriented.

An alternative to such a customer-focused orientation is a production- or sales-oriented company. Production-oriented firms are internally focused on gaining economies of production. Although sales-oriented firms leverage sales skills to "push" products and services to the marketplace, customer-oriented businesses are attuned to and responsive to customers "pulling" the product they want to the market for consumption. Customer-oriented firms also need to work to achieve quality and cost-effective production and sales; but the methods to achieve this must support the customer orientation strategy.

Market-oriented firms require market-oriented personnel. This internal market orientation focuses on creating a customer-focused internal environment well equipped to better provide for the external customer. Figure 5.1 illustrates six core elements necessary for creating an internally market-oriented workforce. External customers are positioned within the center of the diagram to illustrate that their wants and needs are the ultimate center target. Warehouse personnel survey- and interview-based research conducted by the authors supports the need for managers to leverage these six keys to frontline engagement. Doing so can produce customer-centric managers and customer-oriented frontline warehouse personnel. As the diagram illustrates, a failure to do so produces varying levels of managerial effectiveness. Data indicate that only 23 percent of the warehouse employees surveyed felt their managers were fully customer-oriented.

Baton drops resulted for both the men's and women's U.S. sprint relay teams while competing in the 2008 Beijing Summer Olympics. Both U.S. teams were considered to be strong competitors, but the hand-off between an incoming runner and an outgoing runner for each team failed the baton exchange. In a track relay the incoming runner has the ultimate responsibility to place the baton in the proper position/location where the next runner (outgoing) expects the baton to be. This allows the outgoing runner to focus on efficiently beginning his leg of the race. Both runners have unique responsibilities that must be executed properly for the baton exchange to be successful.

Why didn't the proper baton exchange result occur? Often the same question is asked by managers and supervisors when exchanges of information, products, or services fail to execute properly between employees and departments within the workplace. In warehousing and distribution there's an old saying "If the paperwork don't move, the freight don't move." It pertains to information exchanges that are required so that the next-in-line person or operation may take place. Delays in information exchange often result in delays in product movement.

Internally market-oriented employees understand the need to perform their duties and activities successfully so that the next-in-line "internal customer" relying on the output can successfully achieve her task in the process. In a perfect "relay" or process, no employee in the chain of activities should look backward to locate the necessary inputs. The internal supplier should already have their activity's output in the expected location to facilitate an efficient and effective hand-off to the internal customer.

Although every business person understands that not all exchanges will successfully be executed, perfect execution measures certainly should be designed within every work process and imbedded within the minds of every customer-oriented employee. For the warehouse, six primary factors must be executed to facilitate a customer-focused workplace: knowledge development, information exchange, assistance, performance feedback, affirmation, and interdepartmental service orientation. The following sections describe each dimension, and examples from the field illustrate internal customer orientation in action.

## Keys to Frontline Engagement

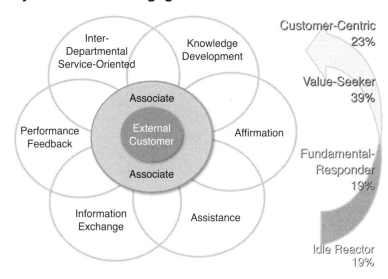

Figure 5-1   Keys to frontline engagement

## *Knowledge Development*

Warehouse employees must learn the mechanical skills required to perform their job tasks. In addition to the basic skills of the job, frontline warehouse employees need to understand how their duties and the output from those tasks impact the value for the co-worker receiving their output and the impact their output has on the overall process. Managers must help to provide both basic knowledge training and higher-level knowledge development.

Basic math and writing skills are necessary for counting product and communicating with co-workers and technology. Take for example, the quick counting of a pallet of product. In industries, such as the consumable package goods business, full pallet quantities of a single product may have a specific "tie-high" configuration. Each tier of product will contain cases positioned according to their dimensions and in a way that when pivoted in the next tier will act as a locking mechanism from one tier to the next throughout the pallet. A clerk may identify that the tie, or the configuration of one tier, contains 10 cases and that there are 5 tiers high of the product on the pallet. In this example, the tie (10 cases per tier) multiplied by the high (5 tiers) would equal a total pallet quantity of 50 cases. Each industry may have a set of standard configurations for each SKU produced and distributed. Knowing the tie-high configuration equips warehousing personnel with the critical information to accurately and efficiently count a full pallet of a single SKU.

Some of the core basic skill requirements for operations warehouse employees include (Lemay et.al 1999, p. 95):

- Basic math and reading
- Safe package handling techniques
- Computer interfacing skills
- Machine operation skills and licensing
- Communication and interpersonal skills
- Basic knowledge of transportation and inventory

Customer-focused warehouse firms offer employees critical training and instruction for performing their jobs but also provide opportunities to advance employee knowledge and skills beyond the fundamentals of the work. One manager had difficulty understanding why a customer service representative would perform his duties in perfect sequence (from most time-sensitive to least time-sensitive) some days, but other days the most important and time-sensitive work was left undone and required unplanned overtime to complete. The manager just thought that the employee simply cared about doing the easy work, but didn't mind doing the hard work with overtime pay.

In the end, it wasn't that at all. No one had taken the time to teach the employee how to identify the most critical tasks of the day and prioritize them at the top of the employee's to-do list. Further explanation was given to the employee as to why one task is more critical than another, and how the delay or failure of the task would impact the awaiting employee and entire process. The employee now understood the importance of his duties beyond the basics of the job. He began performing so well that eventually he moved into a customer service supervisory role.

By helping to advance employee knowledge, companies invest in the employee's future within the company and are more likely to instill a healthy allegiance among the operating warehouse employees. The four core dimensions important in developing knowledgeable warehouse employees include technical knowledge and skills development, formal training, informal learning opportunities, and integrative understanding.

The example of technical knowledge to perform the basics of the job is illustrated in the receiving operator's understanding of how to quickly count full pallet quantities of cartons by obtaining the mathematical product of the tie x height. The customer service representative example illustrates a higher level knowledge capability to prioritize tasks based on an integrative understanding of how the employee's work fits in to impact the outcome of the process. Training must be formalized, and the best training programs have designated skilled trainers—not just an appointed worker who performs the job well. Likewise, training must continue through informal learning while on the job and interacting with co-workers. Examples from various warehouses in the field include

- Classroom type training and job shadowing

- Professional trainers

- Mentorship programs

- Cross-training

- Employee teamwork and involvement

Table 5.1 provides a brief example of some organizations that provide various training and educational opportunities for advanced employee knowledge development. This short list is by no means complete and highlights only one or two of the many educational paths offered by the organizations listed and by other professional, governmental, and educational organizations and associations.

**Table 5.1: Examples of Employee Training Options**

| Paths | Offered By |
| --- | --- |
| Certification Programs Conferences | Council of Supply Chain Management Professionals |
| Online Learning | Warehousing Education and Research Council |
| Certification Program | Institute for Supply Chain Management |
| Diploma Program | American Supply Association (ASA) College of Warehouse Management |
| Targeted Classes | OSHA Training Institute Education Centers |
| Degree Programs Executive Education | University of Arkansas—Departments of Supply Chain Management—Walton College of Business |
| Full-time, part-time, online, in-plant, and resident | |

Employees also should be educated in the important factors that ultimately formulate the customer's perception of the warehouse. Getting employees to buy in to the mission and focus of the firm and to better envision how the employee's job helps the firm fulfill customer needs are important for achieving prolonged high-quality employee performance.

## Information Exchange

Frontline supervisors must provide employees with the information necessary to succeed. One warehouse placed a digital counter next to each receiving door. Unloaders could evaluate their own productivity as they unloaded a floor-loaded trailer by hand. An unloaded case placed on a conveyor would pass a barcode reader, and the digital count display would advance by one carton. At breaks, unloaders would compare the number of cases "thrown" to other unloaders scores. Again, healthy competition was positive motivation for the workers.

Warehouse managers must provide information to employees *specific* to the activities performed by an individual employee. The information must be *accurate* and provided in a form that is *easy to interpret and use* on the job. This means that the information must also be *concise* and to the point.

As one of the authors walked past an information board located within the view of passing employees, he asked the warehouse manager for help in interpreting the graphs displayed on the board. The manager had difficulty explaining that the graphs illustrated performance comparisons between the facility and other warehouses within the company's distribution network. The intention was good—to help motivate employees by providing them the information. However, the information was difficult to interpret quickly, and employees seemed to ignore the charts as if the information were irrelevant to them and managing their work duties.

Within the same warehouse, however, the method for exchanging suggestions, questions, and concerns between employees and the warehouse manager was quite effective. A well-displayed locked box was placed where employees could easily insert their communication for the manager. The policy was that the manager would answer the inquiry within 2 weeks. When the manager believed that the inquiry would benefit all employees, the manager's response would be posted to the information board directly over the box. The information was relevant, concise, and timely. The employees liked and utilized the program well.

Some warehouses have the customer service representative (CSR) located on the warehouse floor near the shipping supervisor, clerk, and operating employees managing the CSR's account. Information exchange is quick and accuracy of the information can be easily verified. This makes communicating with the warehouse client more productive, too. When the client contacts the warehouse, the CSR immediately can provide details to the customer.

Examples of internal customer-oriented information exchange in practice include

- Strategic positioning of electronic information boards in the warehouse
- Computer information kiosks for two-way communication with employees

- Employee led meetings
- Start-of-shift information meetings
- Co-location of key personnel managing client account

## Assistance

Four core elements pertaining to assisting employees can help to promote a customer-focused workforce and environment: proactive assistance to employees to help resolve issues before escalation, responsive to employee issues, immediate assistance deployment, and specifically directed to the employee in need.

Managers must be well attuned to sensing the needs of employees even before the need arises. In one motorcycle production plant, employees rotate to perform other jobs so that physical and mental fatigue is limited due to working only a single task day-in and day-out. An employee performs the same task once within a 2-week period. Job rotations or cross-training is also suggested within the warehouse. Proactive cross-sectional job training programs can assist employees in better understanding the core processes important for the success of the warehouse. Cross-learning can help employees manage the work better if another employee requires help or when the warehouse section is a person short for the shift.

Frontline employees rely on the responsiveness of managers to help resolve immediate issues. Although a complaint does not have to be resolved in the favor of the employee, it needs to be resolved in a timely manner, and the employee must be informed as to the reasoning behind the decision.

- Responsive managers are trained in coaching employees for productive outcomes.

- Cross-learning is important for improving employee understanding in the overall work.

- Working supervisors take part in carrying-out activities of daily processes and are therefore close to subordinates when issues arise.

- Highly trained and disciplined employees may be entrusted and empowered to identify problems and carryout solutions without the help of supervision. This is truly proactive planning on the part of management.

## Performance Measurement and Feedback

A highly productive employee consistently completed her daily tasks before the end of the shift. During the last hour of every day she would be relocated in the warehouse to assist another employee in completing her duties. Unfortunately, the added work was not captured by supervision as additional productivity for the helper. In this circumstance, the helper believed that she was punished for efficiently completing her work. Her only desire was to have all her performance for the day recorded and rewarded.

Warehouse employees must believe that their evaluations are meaningful and specific to the jobs they perform. Although task-driven, the warehouse management system must be flexible enough to capture all the activity specific to an individual during the course of a shift.

Performance appraisal systems must not only be job-specific, but they must also be fully understood by the employee. Making employees part of the process to develop performance appraisals can ensure that the correct elements of job performance are monitored and measured; this can also help attain employee buy-in. By establishing and communicating Key Performance Indicators (KPI) used to measure the output of an employee's job, the performance appraisal system can be optimized. Table 5.2 provides a list of key performance indicators.

**Table 5.2: Example Key Performance Indicators (KPIs)**

| EMPLOYEE | MANAGEMENT |
|---|---|
| ■ Order Entry Accuracy | ■ Inventory Carrying Cost |
| ■ Picking Accuracy | ■ Inventory Turns |
| ■ Lines Picked Per Hour | ■ Order Cycle Time |
| ■ Cycle Counting Accuracy | ■ Workforce Utilization |
| ■ Percentage Damage | ■ Shipping Accuracy |
| ■ Cartons/Pallets Received | ■ Order Fill Rate |
| ■ Trailers Loaded | ■ Customer Satisfaction |
| Measure What Matters | |

Performance feedback must continuously be provided to keep warehouse employees informed of their working progress. Visual methods to communicate employee performance can be helpful if displayed near the employee's workstation. When an employee is reassigned to help in another section of the warehouse, methods must be in place to continue to capture the employee's work productivity. Other examples from the field include

- When appropriate, illustrate performance in units and dollars.
- Calculate employee performance so that employees may easily compare their performance contribution across employees, departments, and facilities.
- Formal evaluations should also include the employee's self-evaluation.
- Individuals and teams should be rewarded for exceptional performance.

## Affirmation

Historically, warehouses and loading docks have been characterized as rugged working environments requiring brawny men to muscle freight from loading dock to trailer. With today's progressive warehouse information technology and advances in "smart" machinery, the workplace for many warehouses and distribution centers has changed in dramatic fashion. Managers spend more time strategizing and planning, and employees spend more time adding greater value to the products they service.

With the changing nature of the workforce and warehousing environment, places of employment must be seen by the workforce as being fair, open communication-based, collaborative, and nonthreatening. Job assignments must be seen as fair and comparatively equitable across co-workers. Managers facilitate the perception by fully explaining how and why specific work tasks are assigned. A team-like atmosphere should be instilled that encourages open communication and praise with and between employees.

For one company, employees work in teams that virtually manage themselves. The team works together to determine and approve vacation times for its members. This may seem awkward at first glance, but the team is responsible for performing the work of its members that are absent from the workplace. No substitute employees are provided to a team.

## Interdepartmental Service-Oriented

Research suggests that when one department demonstrates characteristics of internal customer orientation, it also begins to treat other departments as internal customers who receiving its output. With entire organizations forming customer-oriented environments internally among employees and externally with supply chain partners, the companies can be more competitive in the marketplace.

## Results of Frontline Warehouse Employee Engagement

Figure 5.2 illustrates the important outcomes pertaining to an internally market-oriented workplace. It is especially true for service type organizations such as warehouses that don't own the product but that service the customer by way of managing products, information, and costs. The happy face in the figure indicates that more than 57 percent of the warehouse employees responded that they were approximately 68 percent satisfied with how their managers were providing an internal customer-oriented workplace through knowledge development, information exchange, assistance, affirmation, performance feedback, and interdepartmental service. Managers indicated that the more satisfied group achieved higher productivity and safety, and their communication and overall satisfaction while on the job seemed greater compared to the employees that were dissatisfied in the key areas.

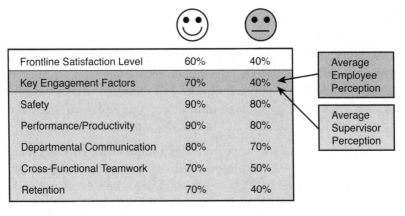

| | 😊 | 😐 | |
|---|---|---|---|
| Frontline Satisfaction Level | 60% | 40% | |
| Key Engagement Factors | 70% | 40% | Average Employee Perception |
| Safety | 90% | 80% | |
| Performance/Productivity | 90% | 80% | Average Supervisor Perception |
| Departmental Communication | 80% | 70% | |
| Cross-Functional Teamwork | 70% | 50% | |
| Retention | 70% | 40% | |

Figure 5-2    Results of internal market-orientation

# Logistics Personnel Development Case Study

The discipline brought to the job by experienced supervisors helps to ensure that employees adhere to established processes that are important to the quality and cost of warehouse operations. Along with discipline, it is also important to establish a rapport with employees that encourages them to communicate and take initiative and pride in their work. The following case study presents a dilemma between three managers in a warehouse. Supervising styles differ significantly between two managers. The tension must be resolved before it impacts the morale of the operating employees and clerks.

## *Logistics Consolidators, Inc. (LCI) (Logistics Personnel Development Case)*

Paul Krass finds it difficult sleeping at night thinking of the relationship and management issues between his co-director and his recently hired college friend. In less than a month, Paul's friend is contemplating tendering his immediate resignation, whereas his co-director wonders why Paul ever suggested the hiring. Paul knew that there would be some personality differences initially to overcome, but thought that in time the two would complement each other's managing style. The situation seems to have come to a boiling point, and Paul is in the middle counseling his friend one minute and his co-director the other. He knows that both managers are valuable to the company and feels that he must help to identify the problems and preventative solutions so that he helps to retain the managers and their friendship.

## LCI

Recently, LCI integrated its operations with Smarter Packaging Services to offer new clients tailored packaging to support its promotional efforts and reduce logistics costs. The arrangement was important and helpful in convincing one new customer, a large manufacture of consumer goods, to partner with LCI for providing logistics services.

Relationship integration with suppliers and customers was a new idea to LCI. The company's 20-year history included regional warehousing, consolidation, and distribution for more than 50 consumer goods manufacturers, and the company's leadership included a husband/wife ownership. The new integrated operation nearly doubles the throughput of inventory; consequently LCI leased an adjoining facility for the new business.

Traditionally, LCI's employee structure included operating employees (lift operators, checkers, inventory clerks, lead operators, and so on), front-line supervisors (managing paperwork flows, worker schedules, and delivery/pick-up driver interaction), account managers (interfacing with shippers and customers, other departments, and supervisors), building managers (interfacing with other managers and directors), and directors (interfacing with building managers, other directors, and executive management, including owners). LCI recently adopted a new warehouse management system and transportation system that helped replace managing product movement and communications through paperwork exchange. In addition, the firm replaced old lift equipment with new units having mobile scanners and smart systems for programming standard lift heights thereby improving employee productivity in order picking and product putaway. The owners thought that the move toward technology would help reduce headcount in the company.

Operating employees and front-line supervisors are typically hired through a temporary agency, and supposing they learn on-the-job well enough, LCI offers them full-time employment. On-the-job training is provided as new hires "shadow" experienced employees throughout the course of a week. Typically, the trainer is an employee that currently works in the area in which the new employee will be assigned. The personnel department places advertisements in local newspapers for LCI's job listings pertaining to account manager and building manager needs.

A typical order proceeds through the following process when product is scheduled for shipping. The order is received by the customer service department and scheduled for consolidation and shipping by the transportation department. Each afternoon pick tickets and bills of lading are sent to the operations department for processing. These orders are picked during the evening and shipped the following morning. Lift operators pick and stage orders, and checkers double-check the product counts. Carriers are appointed a time for pick-up and are expected to count and stretch-wrap the freight. Depending on the consistency of the orders and the experience of the lift operator, a driver may be asked to assist in loading.

# Situation

Paul Krass, Andy Mets, and Shelly Anderson get together Thursday nights to play ping pong, relax, and casually discuss business within their logistics company. Paul is the director of transportation and customer service, Andy heads the distribution center operations, and Shelly is the VP of marketing for Logistics Consolidators, Inc. (LCI).

Paul, a recent MBA graduate, suggested a friend of his from graduate school to be Andy's new operation manager of the new packaging services and the new large client's distribution needs. Paul's friend, Keith, had received top academic honors and had extensive experience running his own business in years past. Andy thought it was a good idea to give Keith serious consideration for the job, especially because a new operation would mean the need for a "self-starting" manager with an entrepreneurial spirit, capable of managing with great amounts of autonomy. It was a bonus that Keith had an advanced college degree.

During the interview, Andy provided Keith a formal job description. The description was obtained from the files of the personnel department, and was efficient because it was kept to one page. Keith and Andy talked for an hour, and after walking through the new facility and discussing an acceptable salary, the decision was made. Keith would start on Monday.

Problems began for Keith the first day of hire. Over the weekend, Keith decorated his office, which included hanging his degrees on the wall. Monday morning Keith had a schedule planned for reviewing his workers' personnel files and personally sitting down and talking with each worker about their job, ideas, and interests. Keith had ideas to evaluate his employees' training programs, processes, and get to know something about them personally. Andy's plans were different. Upon walking into Keith's office, Andy mentioned that he had a degree from "the school of hard knocks," and that Keith should hit the ground running.

After a week, Keith came to Paul for advice. The conversation went something like this:

Keith: Hey Paul, can I talk with you in confidence?

Paul: Sure Keith, you know you can trust me.

Keith: I'm not real sure this job is for me.

Paul: What do you mean?

Keith: It seems that Andy's military service background and managing style conflicts with my ideas of how to treat people. Every time I ask a question, Andy says that if he has to explain everything to me then he'd just do it himself. I've heard him say that to my workers, too.

Paul: Andy is just a hard worker and has been in this business a long time. You can learn a lot from him by just following him around.

Keith: But that's not what I was expecting, following him around. I was hoping for a little formal training and some time to get up to speed. Does he treat all his new-hires like this?

Paul: Like what?

Keith: For example, he makes comments like "employees don't do what they're supposed to do because they don't care." I know he's referring to me, but as I see it Keith doesn't explain all the little details about things, and when I ask him to help me understand he simply says, "I'll be glad to help you with that, but I'm too busy right now."

Paul: Andy is a little direct in his approach, but it does seem to work with these types of workers.

Keith: What "type" are you talking about?

Paul: You're right, distribution center workers today are not the same as yesterday.

Keith: I'll say, with today's technology workers don't have to mess with so much paperwork, and we can expect higher level critical thinking and decision making than we expected even from the supervisors of yesterday. In fact, I suggested to Andy that if we retrained our workers and re-engineered our processes, we might reduce headcount in the supervisor ranks.

Paul: What was Andy's reaction?

Keith: He said that would never happen because workers need constant looking after and double-checking because they are not loyal to companies. He said, "They're just loyal to a paycheck," and that's what he uses to motivate them. I mean threaten them. Listen, when employees hear that they should do something because they are paid to do it and they should do it for the good of the company, all they hear is "they should do it for the good of the owners."

Paul: Hang in there Keith, and I'll talk with Andy.

Very shortly after Keith and Paul talked, Andy came into Paul's office and the conversation went like this:

Andy: Hey Paul, can I talk with you in confidence?

Paul: Sure Andy, you know you can trust me.

Andy: I'm not real sure this job is for Keith.

Paul: What do you mean?

Andy: Well, I tell him things to do that pertain to his responsibilities, but it's like I'm talking in a different language. It seems like I have to hold his hand on everything. He thinks that we should get close to the workers, and I keep telling him that they won't respect you unless you're stern with them and direct them in their every move. I really didn't expect to have to do that to him, too.

Paul: Don't you think Keith will settle into the position in a few weeks?

Andy: Maybe, but I'm not sure I'm comfortable with his style of managing. He wants to bring the workers into planning meetings and include them in decision making. I know most of these people are good lift operators, but I'd be worried about putting the customer in their hands. I told him that we already have policies in place to make the employees feel good. You know we have a suggestion box and an open-door policy. Nobody even uses these.

Paul: Keith has some good qualities and brings some new management techniques with him that should help us in better integrating with our customers.

Andy: Sure, but do we have to integrate with our employees, too? He's talking about internal customers, using the same tools we use for marketing to customers and market jobs to employees, creating a learning organization environment, and training managers and employees in being sensitive to the voice of others and responsive to others.

Paul: What would you suggest?

Andy: First, we pay employees and that should be enough. Beyond that we have integrated our departments and made them more collaborative.

Paul: How do you know we're really any more integrated and collaborative than we were in the past? We still have departments, and our employees continue to be evaluated based on productivity measures specific to their jobs rather than to processes as a whole. Yes, we have quality teams, but how do we know they're working the way they're supposed to be working?

Andy: Now you're starting to sound like Keith. Hey, I know they're working because of the bottom line. When it's good then we're integrating and collaborating.

Paul: Give Keith another few weeks. In the meantime, I'll have a talk with him.

## The Immediate Challenge

Realizing Keith may quit any day, and at best Andy will put up with it another 2 weeks, Paul shut his door and put his head on his desk. How could he come up with a plan to show both managers that they each have things to offer the company, especially in light of the new partnership with Smarter Packaging, the new customer, and the new logistics labor environment? Without learning from each other, they may jeopardize the new business.

## Questions for Discussion

1. What issues are truly important for LCI to successfully service its new business?

2. Under the current management philosophy/culture what results do you expect from the new integrative relationship with Smarter Packaging? How would you approach the relationship?

3. What merits do you place on Keith's suggestions and management approach? How would you measure the success of his management approach to that of Andy's?

4. What suggestions could you make to employ the pros and reduce the drawbacks of each management style while meeting the objective of the owners to reduce headcount?

5. What steps would you take to ensure that operations employees perform to their capability, and how would you improve the internal integration among employees?

6. What did Keith mean by internal customer? Is there any way to measure the outcome of treating employees as internal customers? Is there any way to measure the outcome of internal integration and collaboration?

7. How does managing today's workforce differ from managing yesterday's labor pool?

## Summary of Key Points

- An otherwise strong supply chain can be rendered impotent if warehouse operations—frequently the last link in the chain—are weak.

- Trained and motivated personnel are the salient factor in the success of warehouse operations.

- Managers must be technically sound in warehouse operations, must be well versed in information technology usage, and must be capable of managing people.

- Personnel assignment and equipment utilization can greatly impact the ability to meet budget.

- Performance appraisal systems must not only be job-specific, but they must also be fully understood by the employee.

# Key Terms

- Bill of Lading
- Capacity Utilization
- Counseling
- Customer Satisfaction
- Equipment Utilization
- Floor Loading
- Inventory Accuracy
- Inventory Carrying Costs
- Inventory Control Clerks
- Inventory Management Systems
- Inventory Turns
- Job Shadowing
- Key Performance Indicators (KPI)
- Order Cycle Time
- Order Entry Accuracy
- Order Fill Rate
- OS/D Clerks
- Outbound Shipping Clerk
- Overages, Shortages, and Damages (OS/D)
- Pallet Jacks
- Perfect Order
- Pickers
- Picking Cycle
- Product Handling
- Personnel
- Receiving Clerk
- Reserve Area

- SKU

- Supply Chain

- Supply Change Management

- Tie-high Configuration

- Warehouse Management Software

## Suggested Readings

Autry, C. W. and A. R. Wheeler, (2005), "Post-hire Human Resource Management Practices and Person-organization Fit: A Study of Blue-collar Employees," *Journal of Managerial Issues* Vol. 17, No. 1: pp. 58–75.

Autry, C. W. and P.J. Daugherty (2003), "Warehouse Operations Employees: Linking Person-Organization Fit, Job Satisfaction, and Coping Responses," *Journal of Business Logistics*, Vol. 24, No. 1: 171–197.

Ellinger, Alexander E., Andrea D. Ellinger, and Scott B. Keller (2006), "Supervisory Coaching in a Logistics context," *International Journal of Physical Distribution and Logistics Management*, 35 (9): 620–636.

Ellinger, Alexander E., Scott B. Keller, and Ayşe Banu Elmadağ (2010), "The Empowerment of Frontline Service Staff in 3PL Companies," *Journal of Business Logistics*, 31 (1): 79–98.

Keller, S. B. (2007), *Managing the Functions: Personnel*, Chapter 16, in *Handbook of Global Logistics and Supply Chain Management*, Sage Publications, editors John T. Mentzer, Matthew B. Myers, and Theodore P. Stank, University of Tennessee: 273–282.

Keller, Scott B. Kimberly Hochard, Thomas Rudolph, and Meaghan Boden, "A Compendium of Multi-Item Scales Utilized in Logistics Research (2001–2010): Progress Achieved Since Publication of the 1973–2000 Compendium," (2013) *Journal of Business Logistics*, Vol. 34, No. 2: pp. 85–93.

Keller, S. B. and J. Ozment (2009), "Research on Personnel Issues Published in Leading Logistics Journals: What We Know and Don't Know," *International Journal of Logistics Management*, Vol. 20, No. 3: 378–407.

Keller, S. B., Voss, M. D. and Ozment, J. (2010), "A Step Toward Defining a Customer-centric Taxonomy for Managing Logistics Personnel," *Journal of Business Logistics*, 31 (2): 195–214.

Lemay, S. A., J. C. Carr, J. A. Periatt, and R. D. McMahon, Jr., Mississippi State University (1999), *The Growth and Development of Logistics Personnel*, Council of Logistics Management now Council of Supply Chain Management Professionals, Oak Brook, IL.

Periatt, J. A., S. Chakrabarty, and S. A. LeMay, (2007), "Using Personality Traits to Select Customer-Oriented Logistics Personnel," Transportation Journal, Vol. 46, No. 1: pp. 22–37.

Richey, R. G., M. Tokman, and A. R. Wheeler, (2006), "A Supply Chain Manager Selection Methodology: Empirical Test and Suggested Application," *Journal of Business Logistics*, Vol. 27, No. 2: pp.163–190.

# 6

# WAREHOUSE NEGOTIATIONS, AGREEMENTS, AND CONTRACTS

## Introduction

This chapter discusses the role that contracts have in warehousing services and the process involved in attaining a contract for warehousing services. You explore the tools that put a contract in place and the components within those tools.

## Role of Contracts

All third-party warehouse agreements should be governed by a contract, whether a public or a contract type of warehousing approach is used. Contracts are designed to protect both the user and provider, and should facilitate a healthy partnership between the parties. Contracts codify the expectations of both parties, provide protection to both parties in the event of breach, and establish the foundation for any legal recourse actions if needed. Precontract negotiations should focus on defining the specific requirements and ensuring that both parties understand and agree to the makeup of the cost components so that it is clear how charges are defined and incurred and so it is clear what services are included in the charges. There will be enough operational issues to resolve during execution that are not anticipated initially, so it is advisable to include all service and cost aspects that are known at the time of negotiations and contract emplacement.

# Request for Information (RFI), Request for Proposals (RFP), and Request for Quotes (RFQ)

After requirements have been defined, step one in the process of selecting a warehouse provider is to obtain information about the potential providers of warehousing services in the locations wanted by the user. This is accomplished by drafting a list of potential providers that have warehouse operations in the area wanted. Information may be obtained through industry association websites such as the Warehousing Education and Research Council (WERC.org) and the Council of Supply Chain Management Professionals (CSCMP.org). The area chamber of commerce may also have names and contact information of warehouse providers.

The objective at stage one is to obtain initial information about the potential candidates of warehouse operators. A request for information (RFI) is sent to each potential warehouse. In the RFI packet, the warehouse is informed of the user's intention to hire a warehouse services provider, and a brief description of the warehouse activities that will be required is provided to each candidate. Ideally, a Statement of Work (SOW) is provided in the RFI to provide more specific mission requirements to the warehouse operator thereby facilitating a more complete response. Following is a list of common components contained within the SOW. However, a SOW can be deferred until the Request for Proposal (RFP) stage:

- Description of services required
- Period of time
- Price structure
- Equipment requirements
- Licenses and certifications required
- Deliverables
- Additional technical requirements
- Reporting requirements
- Inventory control system requirements
- Personnel qualifications required
- QA/QC plan requirements

The respective warehouse operators then assemble an information packet specific to the services of each warehouse. Information should include services offered, size and condition of the warehouse, available equipment and personnel, type of products handled, license and operating information, and certifications and credentials along with any other operating information that could be helpful to the user when making an initial down select. Contact information should be included, and the response to the RFI should be submitted by the deadline indicated in the RFI call. Failure to submit on time can be reason for disqualification from consideration.

After the deadline for RFI submission passes, the user reviews submissions to identify potential providers that seem to have the capabilities and facilities to effectively offer services required by the user. These providers are down-selected, and the other submissions are set aside as the user focuses on the potential providers. Stage two begins with the announcement and release of a Request for Proposal (RFP). In the RFP, users provide potential warehouse service providers with more detailed information about the service and space needs. The SOW is updated, using information gained during the RFI stage, and within the request, the user provides specific product information including product size and weight, number of SKUs and anticipated velocity, number and size of orders, and special handling needs which may include temperature controlled product or value-added services wanted. The RFP should also provide the potential warehouse providers with the criteria that can determine if respective bids are worthy of moving to the Request for Quote (RFQ) stage. Evaluation criteria may include elements such as overall experience, past performance providing the specifically stated warehouse services, industry certifications and quality assurance programs, and the estimated price to perform the stated services.

The user provides specific product and service information to allow potential providers the opportunity to evaluate their capabilities and interest in continuing on to the proposal and negotiations stages.

In the proposal response (bid) to the RFP, providers share with the user more specific information about their ability to handle definitive product volumes and propose plans to manage the potential client's business. Proposals should be specific to the expected volume, service expectations, and equipment and labor required to perform the operations. The submitting warehouse will also provide cost estimates and professional references at this point. Bidders must ensure that they address all the user's RFP stated requirements and respond to each of the stated evaluation criteria to optimize chances of being down-selected to enter the next stage of the process.

Providers making it past the RFP stage would then be asked to submit specific and final pricing to the potential client in response to a RFQ. This would be the stage in the process leading up to final negotiations. Fees should be communicated according to the units wanted by the client. For example, a client may be interested in evaluating costs at the pallet level including receiving, putaway, storage, picking, and loading. Another potential client may want a per-case cost including all the handling charges and then the storage charges per case quotes separately. It is important to convey pricing information in a straightforward and easy-to-understand manner. This enables ease of evaluation by the user and helps ensure that there are no misunderstandings as to the price components. Users may request that bidders provide information on their price realism and reasonableness in comparison to the established/specific market. This type of information provides the user qualitative aspects of market pricing to use along with the quantitative pricing data submitted in comparing quote submissions.

When all RFQs are received and evaluated, the user may decide to set up interviews with the final candidate warehouse providers to further discuss and explore the services and associated costs prior to making a final selection decision. A savvy warehouse client will conduct its own independent cost estimate to use in comparison to the price submissions made by bidders. This independent cost estimate serves several functions including assists the user in fleshing out its requirements, provides a baseline to evaluate bidder pricing submissions, and provides a platform for use during discussions and negotiations. It is during the negotiations stage that costs are discussed in detail, and agreements on responsibilities and expectations are finalized. Selections of primary and secondary warehouse providers should be decided immediately after formal negotiations are conducted.

## Negotiating

By the RFQ stage details are provided on service capabilities and pricing necessary to begin final negotiations. Potential providers have been reduced to the final set of candidates for consideration. Warehouse operators are experts at evaluating client needs and estimating personnel and equipment costs to successfully provide for clients. Clients are advised to conduct their own service and pricing analysis prior to entering into the negotiations stage. They should be well equipped with information and analysis so that the outcome of the negotiation will be successful for the user and the provider.

## Contract Sections and Content

Contracts should be tailored to the client and provider specifications; however, there are sections that should be included in all contracts. Templates for Warehousing Services Agreements/Contracts are available from various sources, including the U.S. Government General Services Administration (GSA), to use as a starting point. At a minimum, contracts should include enumeration of the warehousing services to be provided, liability provisions, payment terms, and delivery requirements.

Following is an example Warehousing Services contract for reference purposes.

## Example Warehousing Services Agreement

State the legal names of the two entities entering into the agreement and the location.

For Example: This agreement is entered into as of the XX day of XXX, 20XX by and between the Client, and the Warehouse Company organized in the state of XXXX.

1. **Term of the Agreement and Termination Provisions**—Include the dates that the agreement begins and ends. Include the provisions that allow either party to terminate this agreement and the advanced notice required.

2. **Services to Be Provided**—Define all the services that will be provided; include all value added services. Include any restrictions or requirements on physical location of client's goods within the warehouse.

3. **Description of Goods to Be Stored**—State or clarify any storage and/or handling characteristics of goods to be stored in terms of health, safety, or environmental requirements. Clarify terms under which the warehouse may refuse acceptance of inbound freight. State whether stored goods will be subject to any financial liens.

4. **Liability Provisions and Limitation**—State the amount of liability that the warehouse will be held accountable for or limited to. Clarify any conditions that will negate liability.

5. **Insurance Obligations of All Parties**—Define the types and amounts of insurance that the client will be required to attain and maintain. Include all indemnity provisions. Define the type and amount of insurance that the warehouse will be required to attain and maintain to include warehouse employee insurance requirements.

6. **Force Majeure Provisions**—Clarify responsibilities and liabilities in the event of damage or disruption of service due to an act of God, fire, flood, war, civil disturbance, interference by civil or military authority, or other causes beyond their responsible control of the parties.

7. **Confidentiality Agreement**—Define what information will not be disclosed outside of the agreeing parties. Define the length of the confidentiality period. Define penalties for violation of confidentiality agreement.

8. **Noncompete Restrictions**—Define any competitive restrictions to include warehouse prohibitions from storing like goods or goods of a competitor.

9. **Warranties**—These can include provisions warranting condition of goods to be delivered, accuracy of count, labeling, and quality of goods to be delivered.

10. **Dispute Resolution Provisions**—Define the intended and agreed upon manner that any disputes will be handled. Specify the geographic area of jurisdiction.

11. **Rates and Charges**—Define the rates and charges to be paid for services provided. Include the terms of payment such as payment due date/frequency, interest penalty for late payments, and liability for collection costs incurred.

12. **Rate Schedule**—Defining the service to be provided, the time periods associated with it, and the costs for same. Services can include storage, handling for receiving and shipping, transportation services, kitting, special packaging, staging, and minor assembly. Clarify if there is a minimum/maximum service charge.

## *Example Warehouse Receipt: Terms and Conditions*

1. **Ownership of Goods**—This section should include a statement certifying the client ownership status relative to the property being stored. Include a statement relative to indemnification by the client in the event of ownership challenges.

2. **Services Provided**—The section should define all services provided.

3. **Rates and Charges**—This section should include applicable information from the contract Rate Schedule.

4. **Payment Provisions**—This section should state the payment amount as per the rate schedule.

5. **Warranties Agreed to by the Client**—Include client warranties regarding goods to be delivered: accuracy of the count, quality of the content, and so on.

6. **Lien and Security Interest**—Define lien provisions and any applicable security provisions or charges.

7. **Corrective Action Notice**—Define the time period in which notice must be given about inaccurate or incomplete deliveries.

8. **Relocation and Termination**—Define the circumstances under which stored goods can be moved for operational purposes and advanced notice requirements. Define default actions authorized and the conditions under which these will be taken.

9. **Warehouse Liability**—Define the expectations of the warehouse in terms of preventing damage, deterioration, or loss of goods stored. Define the limits of liability for failure to uphold expectations. If applicable define liability calculations.

10. **Insurance**—Clarify insurance provisions and whether the total payment amount includes any insurance amount.

11. **Change of Address**—Define the notice requirements for relocating.

12. **Claims**—Define the claims procedures. Define the notification time period for making a claim. Define any agreed upon alternative dispute resolution process.

13. **Legal Jurisdiction**—Define the geographical jurisdiction for any litigation or arbitration.

14. **Severability**—Define the conditions justifying the cancellation of the agreement.

15. **Signatures Required**—Define original signatures required. Clarify whether electronic signatures are acceptable.

The process for attaining warehouse support, as described in this chapter, is summarized in Figure 6.1.

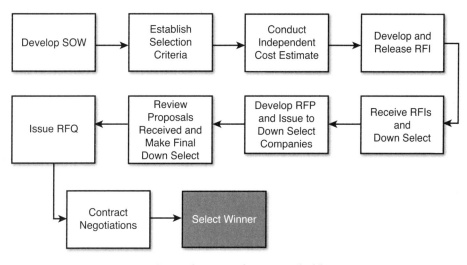

Figure 6-1    General process for service bidding process

## Summary of Key Points

A contract is the mechanism that establishes the business relationship between a user and a warehouse services provider. The contract ensures that both parties understand the requirements and it provides legal recourse in the event of a claimed breach of the agreement. The process of selecting a warehouse services provider is iterative in nature and contains steps and stages that ensure the user requirements are conveyed and understood, the providers capabilities are explained, and the price of providing the required services is agreed upon.

## Key Terms

- Bid
- Contract
- Evaluation Criteria

- Council of Supply Chain Management Professionals (CSCMP)
- General Services Administration (GSA)
- Independent Cost Estimate
- International Standards Organization (ISO)
- Negotiation
- Price Realism
- Price Reasonableness
- SKU
- Quality Assurance/Quality Control (QA/QC) Plan
- Request for Information (RFI)
- Request for Proposal (RFP)
- Request for Quote (RFQ)
- Statement of Work (SOW)
- Value Added Services (VAS)
- Warehousing Education and Research Council (WERC)Warranty

## Suggested Readings

Halldórsson, A. and T. Skjøtt-Larsen, (2006), "Dynamics of Relationship Governance in TPL Arrangements a Dydactic Perspective," *International Journal of Physical Distribution and Logistics Management*, Vol. 36, No. 7, pp. 490–506.

Logan, M. S. (2000), "Using Agency Theory to Design Successful Outsourcing Relationships," *International Journal of Logistics Management*, Vol. 11, No. 2: pp. 21–32.

Lukassen, P. J. H. and Wallenburg, C. M., (2010), "Pricing Third-Party Logistics Services: Integrating insights from the Logistics and Industrial Service Literature," *Transportation Journal*, Vol. 49, No. 2: 24–43.

Lynch, C. F. (2004), *Logistics Outsourcing*, 2nd ed., CFL Publishing, Memphis, TN.

Olander, M. and A. Norrman, (2012), "Legal Analysis of a Contract for Advanced Logistics Services," *International Journal of Physical Distribution and Logistics Management*, Vol. 42, No. 7: 673–696.

Poppo, L. and T. Zenger (2002), "Do Formal Contracts and Relational Governance Function as Substitutes or Compliments?" *Strategic Management Journal*, Vol. 23, No. 8, pp. 707–725.

# 7

# WAREHOUSE MANAGEMENT

## Introduction

Process management is the essence of successfully managing warehouse operations. Mapping out the dynamic flow of goods and the accompanying value-added services (VAS) that occur in an active warehouse is the first step to attaining management success. Eliminating process variations and ensuring full employee understanding of the process—particularly their specific portion of the process—is imperative. Establishing and tracking metrics along with the integration of available technology can enhance the efficiency of warehouse operations by enabling continuous improvements to processes.

### *Process Management in the Warehouse*

Managing the flow of goods and value-adding activities requires employees and management having skills in process management. All primary warehouse processes are a compilation of multiple microprocesses and activities. Outcomes of each activity and process must be articulated and measured in terms of expected time, units, quality, and cost. Employees should be well trained in understanding and achieving the outcome measures. In addition, employees should understand how their performance in each activity impacts the overall process and integrates into the overall suite of business processes.

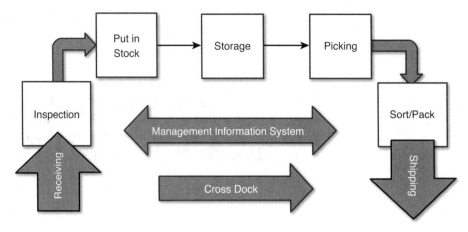

Figure 7-1 *Example of warehouse process flow*

## Variations in Processes

Activities within a process should be well understood by the employees responsible for them. Processes are composed of activities that when managed properly result in efficient and cost-effective operations. Variations in performing activities lead to variations in quality of outcomes, processing time, and cost. For these reasons processes must be simplified and contain only the activities that add value to the processes. All wasteful activities should be eliminated. When efficient and effective, a process must be standardized and all variations eliminated to be optimized. A process must be well understood by the employees performing the activities within the process.

## Process Mapping

Ensuring that a process meets all the requirements for being effective takes a disciplined approach. Process mapping requires the documentation of all activities in a process and an assignment of the employees responsibility for each activity. Outcome measures (metrics) specific to each activity must be defined and measured. Parameters and upper/lower tolerances of each activity should be determined and assessed for feasibility and acceptability. Acceptable process times and expectations of units to be managed must be identified for each activity, and employees must be trained and equipped to meet those expectations.

Well-designed processes should be visually diagramed to facilitate successful employee training. Process maps should be well displayed for quick reference by employees. They also help to identify activities needing change, thereby, facilitating continuous process improvement. Employees can better understand their integrative roles and how they help to achieve the goals of the firm.

## Bottlenecks or Capacity Constraints

By nature, activities have capacity levels that influence the total output potential of a process. Process capacity is determined by the activity having the lowest level of output. It is this activity that causes a bottleneck or constraint on the process and determines the maximum capacity output for the process. Warehouse managers must identify the capacity constraints within each process. It is only then that a manager may have a clear and realistic performance expectation of the process.

## Sequential and Parallel Processing

Managers and employees, alike, should continuously seek ways to improve processes and improve the capacity availability. Process mapping illustrate the sequence of activities that make up a process. Sequential activities occur one after another. As such, the processing time is the sum of the time required to complete each activity. Activities that may be performed at the same time are processed in parallel. It is possible to reduce the total processing time by conducting activities in parallel when possible.

# Primary Warehouse Activities and Processes

Many processes are utilized in warehousing to receive, put-away, replenish, pick, stage, and load product. In each of these overarching processes, there are many subprocesses that may vary greatly due to the nature of the work performed for each warehouse client.

## *Receiving and Putaway*

Complex receiving processes may include the following sequential stages. First, an inbound truck driver provides the shipment documentation to security to receive clearance to enter the distribution center property. Receiving door assignment may be provided by security, or a driver may have to park his equipment and enter the office of receiving to obtain a door assignment. After the trailer is positioned at the appropriate door, the receiving personnel unloads product from the trailer and checks for accuracy in content description, count, and quality. After unloading, a driver receives a receipt verification signature by the receiving clerk, and the driver exists the property.

Each of the four stages discussed make up receiving. The first stage may be further broken down into the individual activities required to complete the process:

- The driver pulls into line to gain entry onto property.

- The driver provides documentation to security (bill of lading and appointment verification).

- Security conducts driver and equipment inspection for safety and security purposes, verifies documentation within the warehouse management system, and obtains a system-generated door assignment for the trailer.

- The driver receives a door assignment and proceeds to the door location.

- The driver stops at the door location and opens the trailer doors.

- After safely backing up to assigned door, the driver locks the trailer to the receiving dock and secures the tractor and trailer with wheel chocks.

- The lift operator ensures the dock leveler is in position prior to initiating the unloading process.

Warehouses are managed through an intricate web of activities and processes. After an inbound trailer is positioned at a receiving door, the warehouse personnel take over and perform tasks to unload the product. Although receiving activities may vary from one distribution center (DC) to another, the goal of receiving is to efficiently and safely discharge product from within the trailer, inspect the contents, notate discrepancies in items, count, and quality, physically move inbound product to an assigned slot location, and update the inventory records to indicate the availability of the product in the warehouse.

## Interface with Driver

Truck drivers may interact with multiple warehouse employees. Prior to arriving at the warehouse, a driver may place a call to obtain a delivery appointment. Receiving supervisors and lead clerks work together to establish a manageable schedule of inbound trailers for the shift.

Drivers interact with security personnel upon entering the grounds of the distribution center. Some drivers drop their trailers in a designated area or door and hook to another trailer, empty or loaded, and depart the facility. Drop trailer processes may occur during regular shift hours or after hours. Drivers have limited interaction with receiving personnel under a drop trailer process.

Somewhat opposite of a drop trailer process, a driver may be delivering a load that must be discharged while the driver waits. Live unloads require the driver and the receiving clerk or supervisor to exchange paperwork and signatures to govern the unloading process. Depending on the policy of the warehouse, drivers may wait in their truck cabs or in a designated driver waiting lounge.

Supply chain-oriented and process-driven warehouse personnel understand the importance of facilitating live unloads and getting drivers back on the road. In the same way, a driver must understand the importance of meeting delivery schedules and proactively communicating any delays to the receiving group.

## Unload and Checking

Unloading may require forklifts for palletized freight or conveyor equipment for floor loaded trailers. Either way, freight must be removed from the trailer, inspected, and counted. Only then can it be placed into available inventory and the information system updated. Receiving dock space must be allocated for the receiving activities that include unloading a trailer and staging the inbound product on the dock. Only after product samples from the inbound trailer pass intensive quality inspections will the product be put away and the driver receive a clear delivery receipt. Inspections may be as simple as looking at the outer cartons to identify any problems with creasing, crushing, or soiling packages. However, more intensive inspections may entail tests for the durability of the product.

## Improving the Process

Drop trailer programs have emerged to help truck drivers and motor carrier equipment stay in motion rather than sitting idle awaiting unloading. By reducing the dwell time of content and carriage, warehouses have helped to improve supply chain efficiencies. For live unloads, it is time-consuming to count product and verify that what is received accurately reflects the documentation. Some receiving processes have removed that activity favoring the more efficient shipper load and count policy. This is where the shipper of the product accepts responsibility for counting the product and releases the carrier from being responsible for the product accuracy. Risks are minimized for shippers utilizing full truck load (TL) service where shipments are loaded, secured, and transported directly to the final destination without opening the trailer for cross-docking or consolidating additional freight. Shipper load and count agreements significantly reduce the waiting time and liability for a carrier. Reductions in both should lead to reduced costs for the carrier and lower freight rates for shippers.

Technology has helped improve the accuracy of warehouse processes. As more manufacturers and warehouses adopt radio frequency identification (RFID) technology for tracking product throughout supply chains, processing time will be reduced and accuracy will be improved. A study of several consumable packaged goods warehouses revealed that by utilizing RFID tags on products, small reductions in processing time for individual cases equated to significant overall reductions in total process time. Processes should be designed with a focus on continuous improvements that lead to reductions in process time and complexity while achieving greater service levels.

### *Replenishment of Forward Picking Area*

Reserve and active product flow requires refilling or replenishing the active empty slots after picking has occurred. Because it is advisable to keep people and product safe by restricting entry into the reserve area, managers should assign duties of replenishment to specific operators. Designated operators become familiar with product locations and know when to move product from reserve status to active picking areas. Inventory integrity also improves as activity within sections of the reserve product is restricted to the responsibility of only a few workers.

## *Picking and Staging*

Order dynamics, product layout, characteristics of items, and many other factors must be considered when designing a picking and staging strategy. Both require efficient time and space utility to keep operating costs low and service high. Although many combinations of picking strategies exist, there are a few common picking strategies that discussed here: single-order, zone, active/reserve, and batch picking.

## Single-Order Picking

Single-order picking assigns individual orders to a pick operator who then picks one order at a time. In a picking tunnel, for example, an order picker may access the order documentation and begin picking from the SKU assigned to that order nearest the picker. A pick-to-light system may be in place that identifies the location of the next nearest SKU with a light indicator. The picker travels to the location of the light and picks the number of units showing on the order and displayed at the pick location. After an order is complete, the box is placed on a conveyor and travels to the shipping dock. Pick-to-light systems vary in design and may involve the use of other mechanical product handling devices for single-order picking. Carrousels, vertical and horizontal, operate by bringing the product to the picker, and a light tree indicates the pick location and number of units to pick. Figure 7-2 illustrates an automated handling system. In a similar manner, robots are utilized in one office products distribution center to bring moveable racks of SKUs to the operator's picking station.

### Picking Techniques

Picking techniques include single-order picking, zone picking, bulk picking, and batch picking.

Figure 7-2    Automated product handling technology (Image courtesy of Baloncici/Shutter-stock)

## Designated Zones

Small consumer orders packaged for parcel carrier home delivery are picked by one ware-house using a zone picking strategy. Operators are assigned to designated inventory pick-ing zones; whereby, each operator is responsible for picking and staging items appearing on each order stored within the operator's zone. Multiple operators may complete the picking of a single order assuming that the order contains SKUs located throughout more than one pick zone. Zone picking allows an operator to gain experience and familiarity in handling a finite set of SKUs and has been shown to improve picking accuracy and speed. Zone picking may also be employed when utilizing pick-to-light systems in the order fulfillment process. The picking lane is divided into connecting sections or zones. When the operator in zone one completes his or her final pick for the order, the pick-to-light system indicates that the order is complete for transporting to shipping or that the order is now ready to move to zone two for continued picking by a different order filler.

## Active and Reserve

As previously indicated, some inventory is particularly applicable for storing in active and reserve locations. Full pallet quantities received and stored in anticipation of future demand may be stored in a section of the warehouse situated away from the picking area that experiences the greatest daily activity. Active picking areas contain product that is immediately available for filling orders, and it is replenished with product from reserve.

## Bulk Pick Line

Orders scheduled for next-day shipping may be considered for bulk picking. Under this strategy, a warehouse inventory management system compiles the total units to be shipped by SKU number. A bulk pick list is generated and stipulates the SKUs' corresponding volumes and sequence of picking that minimizes the picking time and distance associated with all orders to be shipped next day. Single items are picked in bulk and transported to an active picking location called a bulk pick line. Next day's orders are filled directly from the bulk pick line, and at the end of the shipping cycle, the bulk pick line should be reduced to zero. Product remaining in the pick line would be tied to a delayed or canceled shipment. Under the worst case any product remaining could indicate that a miss-shipment occurred. The strategy is a means to isolate problems with overages, shortages, and damages (OS/D) during shipping to a minimal known number of orders and within a specific time frame. It allows for shipping and inventory control to respond to such issues and quickly rectify problems and reconcile inventories.

## Batch

Picking may occur in batches or waves determined by a number of factors that may include product type, location in the DC and volume ordered, mechanical equipment utilized, carrier type, consignee location, and others. For example, a conveyor system's capacity and availability may dictate the sequencing of batches throughout a picking and shipping shift. The distribution center may schedule the first batch to include orders destined for delivery to the furthest retail store locations. Drivers transporting to the distant destinations will require more time to transport and must be loaded prior to trailers assigned to destinations that are nearer the shipping point.

Batch picking may also be scheduled so that picking activity is leveled across employees and sections of the distribution center. Rather than scheduling all units of a high-volume, single SKU for batch one and risk overwhelming the specific pickers and area, the strategy should be to disperse the requirements across multiple batches and moderate the picking demands throughout the entire shift.

Full truckloads, customer pickups, and even parcel package orders also may be important factors for considering batch picking. Managers must evaluate the influencing factors of their specific picking operations to determine if batch processing would be beneficial.

## Loading and Shipping

When picked, products must be staged and loaded onto awaiting transport equipment such as trailers, railcars, air freighters, or ships and barges. Picked product may be staged in the order that it will be loaded into the trailer. The loading order also must accommodate the efficient unloading sequence by destination.

## Checking

Outbound orders must be checked for accuracy and quality prior to loading. Both should be checked during the picking process, and in some operations product is picked and directly moved onto an outbound trailer. This is especially effective when full pallet quantities of product are ordered and the warehouse is equipped with RFID technology for scanning product as it is loaded. However, even under shipper's load and count strategies, a final check for accuracy is advisable. RFID technology is a progressive means to ensure order accuracy. However, not all warehouses can justify the fixed and variable costs associated with the technology. Clerks may then be assigned to check the freight one last time prior to loading.

## Loading

Warehouse and shipper agreements determine the responsible party for loading transport equipment. Truck drivers ultimately are responsible for the legal weight limits, distribution of their loads within the trailer, and the safety/security of the product during carriage. Oftentimes, truck drivers and helpers (sometimes called lumpers) work together to load freight into a trailer. Lumpers are casual laborers hired on a case-by-case basis to assist drivers in the loading and unloading of product, especially when manual labor is necessary to hand carry and stack cases of product.

Other times, trailers are loaded by warehouse personnel. This may entail utilizing forklifts or pallet jacks, or may require floor loading trailers so that all the available trailer capacity is efficiently utilized. This helps to spread the cost of the transportation across a greater number of units and thereby reduces the overall total cost for the shipper while gaining efficiencies for the carrier. Moveable conveyors are utilized to bring individual cartons to loading personnel located inside the trailer. Special bracing, the use of moveable bulkheads, or cushioning technology may be necessary for securing and protecting the load during transit. This is managed during the loading process and may require additional materials, labor, and time to integrate into the load.

## Summary of Key Points

- Understanding and managing the flow of goods and the value-added services (VAS) that compose the activities within an active warehouse is key to optimizing the efficiency of warehouse operations.

- Mapping the processes is the first step necessary to understanding and controlling warehouse operations.

- Each element of the process must be broken down into granular steps and analyzed for efficiencies.

- Establishing performance metrics and ensuring that all employees understand the process and the metrics is imperative to success.

- Eliminating process variations simplifies the workers job and enhances the quality and timeliness of the process.

- Managers must evaluate the influencing factors of their specific operations when determining which type of picking processing would be the most appropriate.

- Technology can be integrated into the process to reduce time and increase accuracy and quality. Training employees on the technology capabilities must be a continual priority.

## Key Terms

- Active Product/Picking Area
- Batch Picking Strategy
- Bill of Lading
- Bottleneck
- Bracing
- Bulk Picking
- Capacity Constraints
- Carrousel Picking System
- Distribution Center
- Drop Trailer Process
- Dwell Time
- Inventory Integrity
- Electronic Data Interface (EDI)
- Forward Picking Area
- Full Truck Load (TL)

- Inventory Management System
- Live Unloads
- Lumpers
- Metrics
- Moveable Bulkheads
- Pallet Jacks
- Overages, Shortages, and Damages (OS/D)
- Parallel Processing
- Pick-to-Light System
- Picking
- Picking Strategies
- Picking Tunnel
- Process Mapping
- Process Time
- Process Variations
- Putaway
- Sequential Processing
- Radio Frequency Identification (RFID) Technology
- SKU
- Reserve Product/Picking Area
- Shipper Load and Count Agreements
- Single-Order Picking Strategy
- Staging
- Task Interleaving
- Value-Added Services (VAS)
- Zone Picking Strategy

# 8

# WAREHOUSE PERFORMANCE

## Introduction

This chapter explores the relationship between warehouse space availability and the layout of equipment and product flow. It covers tools and strategies for optimizing space availability and gives an example of warehouse space calculations. In addition to space utilization, this chapter addresses optimizing worker productivity. You see the importance of process analysis and metric selection. Equipment utilization is the third optimizing element this chapter discusses. Mapping the complete order process and measuring key performance parameters is critical to optimizing warehouse operations, as shown in an example of common performance-type measures.

## *Space Evaluation and Utilization*

It was rumored that one warehouse planner designed the warehouse layout utilizing the skylights as markers for the racking system. As luck would have it, the racks were in place before he realized that they were aligned perfectly with the skylights, and when full the product in the racks blocked the natural light from entering the warehouse.

Warehouse space availability is directly impacted by the layout of the racking system and equipment, the products and their flow dynamics, and the planning that goes into the utilization of the warehouse space. Warehouse operators must consider the total space designed into the building, the maximum space available for operations within the building, and the actual space to be utilized to optimize warehouse operations.

Total space within the four walls of the warehouse must accommodate product flow, but also be allocated for administrative offices, record keeping, OS/D, pallets, aisles, racks, and other storage systems. Machinery, battery charging and maintenance, loading docks, personnel, and driver areas, including break rooms and waiting rooms, and other considerations (such as honeycombing needs) must also be accommodated. Honeycombing pertains to the empty slots among the product in the warehouse and is necessary to facilitate the ease of product putaway and picking. Honeycombing can also identify unused space due to smaller pallet quantities occupying a space that is larger than necessary. Warehouse operators must evaluate the need for such safety capacity in advance so that bottlenecks are not inadvertently caused by lack of planning for enough honeycombing.

For example, consider the shipping dock space required for staging product to be loaded on 10 each outbound 53' trailers. Assume that the product weight will not exceed the maximum legal weight for the trailer and that a total of 52 double-stacked 48"X 40" pallets are to be loaded on each trailer. If the night shift picked and staged each load in the exact configuration that it would actually be loaded into the trailer, enough space will be needed to fit 260 pallets on the floor in the staging area. The product would require access aisles to manage the product between storage and staging, and between staging and loading. Aisles would increase the dock space required (see also, Napolitano and Gross Associates 2003, p. 47).

Space utilization ratios can be calculated to provide a number of metrics for evaluating the space utilization of a warehouse. In the previous example, the maximum number of pallets on the receiving dock floor at any given time is 260. Suppose that your carriers and warehouse procedures allowed for the live loading of full pallet quantity orders directly onto one-half of the trailers. Over time, your metrics would indicate that the shipping dock was operating at a maximum of 50 percent capacity because it was designed for 260 pallets, and only one-half were now staged in the shipping area (130 / 260 = 50%). Under the new circumstances management would be advised to consider reducing the shipping dock space and allocate the excess space to a more useful process.

Of course, total available warehouse storage space should be calculated and a total space utilization metric designed (total number of pallets in storage / total number of pallet positions available for storage). The available storage space will not be the same as the designed warehouse space because the designed space includes offices and other functions not utilized for product storage. Depending on the type of SKUs managed and the equipment utilized to facilitate product storage and movement within the facility, it is advised to evaluate the space utilization by product groupings or by specific areas within the warehouse. Space utilization may be calculated by number of units, square footage, and cubic footage.

The most-sophisticated warehouse and inventory management systems can help managers manage cubic space utilization by assigning inbound product to pallet slot locations based on the height of the product on the pallet. In this way, a pallet containing two layers of product would not be placed in a slot large enough for a full pallet consisting of six layers of product. This would gain efficiencies by reducing honeycombing percentage. In any case, however, space utilization must be based on the available space rather than designed space.

In calculating space requirements for palletized product in racks, for example, many factors must be quantified. First, the maximum number of pallets expected at any given period must be estimated. This would allow enough storage space to manage the warehouse at maximum capacity. Factors for consideration would include

- Case size

- Number of cases per tier

- Number of tiers of product per pallet

- Physical pallet size

- Clearance between pallets

- Clearance above each pallet from the rack

- Honeycombing allowances

- Aisle space

- Other specialized racking dimensions may also be required.

Exhibit 8-1 provides a basic example considering an allowance for honeycombing (see also, Ackerman 1997, pp. 92–93; Napolitano and the Staff of Gross & Associates 2003, pp. 67; and Tompkins and Smith 1998, p. 245). Under the assumptions given in Figure 8-1, the total square foot ($ft^2$) needs for the maximum product volume expected at any one time period would be 11,394 $ft^2$. Of course, aisle space and other relevant factors previously listed would also have to be factored in. Assuming the warehouse did have the required space, the actual space utilized could be compared with the 11,394 $ft^2$ to calculate a capacity utilization percentage. Supposing the average square footage usage for the month was 9,760 $ft^2$, the space utilization factor would be approximately 86% (9,760 $ft^2$ / 11,394 $ft^2$). This percentage could be tracked over time to evaluate an increase or decrease in space utilization.

## Exhibit 8.1: Example of Warehouse Space Calculation

*See also, Ackerman 1997, pp. 92–93; Napolitano and the Staff of Gross & Associates 2003, pp. 67; and Tompkins and Smith 1998, p. 245.

**Cubic foot needs for a bay of product:**

[(pallet width + clearance between pallets) (pallet length)] x {[(case height) (number of tiers high of cases per pallet)] + wooden pallet height + space above each pallet for racking system}{number of pallet positions high}

Example:

- Pallet width = 3.33'

- Clearance between pallets = 0.50'

- Pallet length = 4.00'

- Case height = 2.00'

- Number of tiers high of cases per pallet = 3 tiers

- Wooden pallet height = 0.50'

- Space above each pallet for racking system = 0.50'

- Number of pallet positions high = 2

**Given the previous information, the cubic foot requirements for a bay high of product would equal**

[(3.33' + 0.50') (4.00')] {[(2')(3)] + 0.50' + 0.50'} {2} = **215 ft$^3$** (rounded).

**With the inclusion of honeycombing**

(space needs per bay) / [(1-honeycombing space %)(cases per bay)]

In this case assume 19 percent allowance for honeycombing and 32 cases on each pallet.

**Given the preceding information, the cubic foot requirements for a case product and when considering honeycombing allowance would equal**

(215 ft$^3$) / [(1- 19%)(64)] = **4.15 ft$^3$** (rounded).

**The total cubic foot (ft$^3$) storage space required at a maximum number of 40,000 expected cases at any one time is**

(40,000 cases) (4.15 ft$^3$ per case) = **166,000 ft$^3$**

**Maximum pallets expected on-hand at any given time:**

(40,000 cases) / (32 cases per pallet) = 1,250 pallets

**Maximum pallets on the floor of the warehouse expected at any given time:**

(1,250 pallets) / (2 pallets high per bay) = 625 pallets on floor

**Square foot (ft$^2$) needs per pallet:**

(pallet width + clearance between pallets) (pallet length) = (3.33 ft + 0.50 ft) (4 ft) = **15.32 ft$^2$** needed per pallet footprint.

**Square foot (ft$^2$) needs per pallet when allowing for honeycombing:**

(Honeycombing allowance)( ft$^2$ space for one pallet footprint)

(19%)(15.32 ft$^2$) = 2.91 ft$^2$

(ft$^2$ space for one pallet footprint) + (honeycombing)

(15.32 ft$^2$) + (2.91 ft$^2$) = **18.23 ft$^2$**

**Square foot ( ft$^2$)needs for maximum number of pallets at any given time:**

(625 pallets on floor)(18.23 ft$^2$ per pallet considering honeycombing) = **11,394 ft$^2$**

For a public warehouse, a client may estimate an 86-percent space utilization, but in reality utilize only 65 percent of the space available. In this example, the 21 percent reduction in utilization would mean the warehouse operator has almost 2,400 ft$^2$ that could be utilized for other revenue generating work.

## Personnel Utilization

Time and space are important to manage within all warehouse operations. Processes should be broken down into activities, and both should be evaluated for minimum, maximum, and average time for conducting. Process time is influenced by many factors that could include product type, handling equipment, personnel experience, activities or stages in the process, and other factors specific to the product and the space. Similar to creating ratios for evaluating warehouse capacity utilization, managers can also create utilization metrics to evaluate worker productivity in comparison with a task or process standard expectation.

To begin, a manager must evaluate the number of employees required to perform the shift's expected work (see also, Napolitano and the Staff of Gross & Associates 2003, p. 4). Assuming 20 trailers are expected to be unloaded during the shift, the warehouse manager must determine the time expected for an employee to perform each task or activity within the unloading process. Further assume the lift operator removes each pallet from the trailer, stops for a receiving clerk to physically check the product count and condition, moves the inbound pallet to the designated slot location in the warehouse, confirms the location in the information system, and returns to the trailer for another pallet. The time estimate for the process may look similar to this:

1. **Activity 1**—Lift operator enters trailer and removes pallet: 12 seconds.

2. **Activity 2**—Clerk checks product count and condition: 20 seconds.

3. **Activity 3**—Lift operator transports pallet to warehouse slot location and puts away: 60 seconds.

4. **Activity 4**—Lift operator updates information system with product location: 3 seconds.

5. **Activity 5**—Lift operator returns to trailer to continue unloading: 40 seconds.

Total unload and putaway time is 135 seconds or 2.25 minutes. Supposing this is the time standard accepted by management, then the total time to receive and put away the 20 inbound trailers with 52 pallets each would be 2,340 minutes or 39 labor hours. If workers were allowed a 30-minute lunch break and two 15-minute breaks during the day, then 6 (5.57 rounded) lift operators would be required for the receiving activity for the shift.

A receiving operator performance index (ROPI) could be computed utilizing the adopted lift operator receiving and putaway time standard (see also, Napolitano and the Staff of Gross & Associates 2003, pp. 4–5). In this manner, an individual employee's average time for unloading and putting away a single pallet of product during the shift could be calculated and compared with the standard acceptable time per pallet. A result under 1.00 would indicate the employee operating at a level below the standard time. Anything more than 1.00 would indicate the operator was performing better than the standard efficient time level.

In the preceding example, the standard putaway time for a single pallet equaled 2.25 minutes. Suppose a lift operator actually put away 114 pallets during her shift. The employee's 63 percent ROPI would indicate that the productivity of the employee was 37 percent below the standard time expected:

> (Pallets employee unloaded and put away) (Standard time per pallet) / (work time for the employee)

> (117 pallets) (2.25 minutes) / (420 working minutes per shift) = 63 percent.

Performance metrics must be tailored to the work process and activity. It would be unreasonable to expect the picking, checking, and loading process to require the same time standard expected when receiving and putting away product. In the same respect, product type and SKU mix can influence the expected times for handling activities and processes. Managers must work to map processes and establish realistic time standards specific to their operations so that employees may be challenged but not unrealistically overwhelmed. This may be accomplished by collecting timed data for activities and processes performed by the company's employees over time. Employee historical data may also be captured through the warehouse management system employed.

## Equipment Utilization

Heavy lifts and other machinery are utilized in ocean freight terminal operations to position containers, transfer imported yachts from their storage cradles to tractor trailers, and to position and move general freight and palletized freight within and around transit sheds. This sort of work can also be viewed as warehousing whether it includes outside or inside freight storage. Equipment utilized in warehouse operations should all be equipped with meters used to register the operating time of each piece of machinery. Managers would monitor the operating times of their equipment to ensure that each piece is used in the most efficient and effective manner.

One Long Beach, California, ocean freight terminal operator often had yard work and ship work going on simultaneously. Each shift began with a rush of drivers to secure one of the newest utility tractors available at the terminal. Consequently, the oldest equipment would get the least usage when new equipment was available. Under such circumstances, the newer equipment would begin to wear faster, whereas the older equipment was not yet at the required hours to sell or replace according to leasing agreements. Monitoring and comparing the equipment utilization rate for identical pieces of machinery can assist managers in better assigning equipment and leveling the usage rates across machines.

In the previous longshoreman utility tractor driver scenario, drivers were not steady workers at the terminal. Each driver received his job or company assignment from the union hall prior to the shift. A ratio of actual machine operating time divided by the available operating time would help to identify tractors that were over- or under-utilized (see also, Ackerman 1997, pp. 93–94). Comparisons could be made weekly, monthly, quarterly, and annually. For example, for a single daily shift operation, a tractor would have 35 hours available for a 5-day work week. This assumes a driver would have one-half hour for lunch and two 15-minute breaks per shift. A tractor registering 42 hours would be 20 percent over-utilized (42 hours / 35 hours). However, a tractor having 32 hours of operation in the week would be approximately 9 percent below the expected utilization (32 hours / 35 hours).

## Importance of the "Perfect Order"

Achieving the perfect order for each customer would be the warehouse and distribution equivalent to achieving superstardom. Even the simplest order process would require employees to complete each activity perfectly according to the activity's measurement expectations. Although an order process having only five steps would have five individual opportunities for failure, when failures occur within multiple steps, the effect has a multiplying negative impact on customer service. However, the more complex the process for filling orders and the more touchpoints endured by an order often makes the perfect order proposition even more challenging to accomplish.

This brings us back to the importance of mapping the entire order process and adopting meaningful measures for the process and for each activity within the process. Employees well trained in the activities and possessing thorough understandings of their roles within the process will be better equipped to meet the expected quality levels for the outcome measures.

## Critical Performance Measures

As previously discussed, it is important to measure employee productivity levels and the utilization rates of equipment. Measuring the performance of the warehouse requires the evaluation of the many process and work aspects of the operation.

Customers are concerned with receiving their order complete, accurate, damage free, and on time. One confections client refused to allow the warehouse to ship an incomplete order. Therefore, an order that was held because inventory was not available to fill the order was also late shipping. Annual evaluations of the client's warehouses entailed ranking each warehouse in the distribution network based on inventory availability and on-time shipping. When an order missed a ship date, the incident and reason for the delay was documented. The category of Unavailability of Inventory was totaled at the end of the year and compared across warehouses. A more general missed shipment ratio was also utilized to evaluate the warehouses and was determined by dividing the number of orders not shipped on time to the total number of orders for the year.

One major packaged goods warehouse client measured its warehouses based on meeting the required delivery date (RDD) of an order. However, confusion ensued when an order for the client was shipped on time, but the client's receiving customer provided the delivering carrier an appointment date exceeding the RDD.

The receiving department told the carrier that it had enough of the product on-hand that the carrier was delivering, and that it had a promotion item it was focusing on receiving from other carriers at the time. Adhering to the appointment scheduled by the receiving manager, the carrier delivered the order, when it believed to be on time. To the contrary, however, the buyer for the receiving company documented each case as a missed delivery date and reported the annual percentage of missed deliveries to the manufacturer. The warehouse, carrier, manufacturer, and receiving customer met but were unable to resolve the issue and come to an agreement on a common metric to evaluate on-time performance. Second place was achieved by the warehouse; however, it would never receive the distinction as the number 1 warehouse in the manufacturer's network—an accolade that would have paid tangible dividends.

Exhibit 8-2 contains common measurements utilized to assess warehouse and distribution center performance. As previously illustrated, it is important for the internal and external partners affected by the activity and process to take part in designing measures utilized to evaluate performance outcomes. When created, the measures should be agreed upon and placed in writing by the parties.

## Exhibit 8.2: Common Performance-Type Measures
### Performance Ratios

On-time receiving: (number of inbounds received on time / total number of inbounds)

Order-fill rate: (number of orders delivered complete / total number of orders)

On-time shipments: (number of orders shipped on time / total number of orders shipped)

On-time deliveries: (number of orders delivered on time / total number of orders delivered)

Order cycle time: (average time from order receipt to final delivery of order)

Orders shipped complete: (number of orders shipped complete / total number of orders)

Inventory turnover: (cost value of annual units sold / cost value of average inventory)

**Performance total number of exceptions (These can also be converted into average occurrences.):**

- Stockouts
- Backorders
- Delivery receipt adjustments
- Claims
- Damaged cases or orders
- Overages
- Shortages
- Missed ship dates
- Missed deliveries
- Product substitutes
- Returns

### Cost and Utilization Measures

Warehousing cost as a percentage of sales: (average warehousing cost per order / average order in terms of sales dollars)

Warehousing cost per order: (total warehousing cost / total number of orders managed)

Space utilization: (total number of pallets in storage/total number of pallet positions available for storage)

Equipment utilization: (actual machine operating time utilized / total available operating time)

**Other Measures of Performance**

- Timely claim resolution
- Accurate billing
- Timely billing

**Personnel Measures: (These can also be converted into average occurrences.)**

Employee:

Activity or process performance index:

(units employee serviced) (standard time per unit) / (work time for the employee)

- Number of errors
- Number of days absent
- Number of days tardy

Supervisor:

- Number of work days without accidents
- Employee turnover
- Employee complaints
- Employee overtime

## Summary of Key Points

Warehouse space availability is impacted by equipment layout and product flow. Space utilization ratios can be calculated to provide metrics for evaluating utilization efficiency. Tools such as a warehouse inventory and management system can assist in optimizing space utilization within warehouse operations. Maximizing personnel utilization in warehouse operations requires a thorough understanding of the process flow and can benefit from the use of metrics. Equipment effectiveness and efficiency is the third salient element that must be addressed to optimize warehouse operations. Mapping the complete order process and measuring key performance parameters is critical to warehouse operations optimization.

# Key Terms

- Activity or Process Performance Index
- Available Warehouse Space
- Back Orders
- Bottleneck
- Designed Warehouse Space
- Distribution Center
- Employee Productivity
- Employee Turnover
- Equipment Utilization
- Honeycombing
- Inventory Turnover
- Live Loading
- Longshoreman
- Metrics
- Missed Shipment Ratio
- On-Time Delivery
- On-Time Receiving
- On-Time Shipment
- Order Accuracy
- Order Cycle Time
- Order Fill Rate
- Orders Shipped Complete
- Overage
- SKU
- Overages, Shortages and Damages (OS/D)
- Personnel Utilization
- Process Time

- Public Warehouse
- Receiving Operator Performance Index (ROPI)
- Required Delivery Date (RDD)
- Shortage
- Space Utilization
- Standard Putaway Time
- Stockouts
- The Perfect Order
- Warehouse and Inventory Management System
- Warehousing Cost as a Percentage of Sales
- Warehousing Cost per Order

## Suggested Readings

Ackerman, Kenneth B. (1997), *Practical Handbook of Warehousing*, 4th Ed., Chapman and Hall, New York, NY.

Napolitano, Maida and the Staff of Gross & Associates (2003), *The Time, Space & Cost Guide to Better Warehouse Design*, 2nd Ed., Distribution Group, New York, NY.

Tompkins, James A. and Jerry D. Smith (1998), *The Warehouse Management Handbook*, Tompkins Press, Raleigh, NC.

# 9

# THE ROLE OF INDUSTRIAL PRODUCT PACKAGING

## Introduction

Packaging plays a key role in protecting product to ensure that its gets to the customer in the operating condition expected. The protective function of product packaging serves to protect the product from the environment and protect the environment from the product. However, this is not the limit of the role that product packaging plays. As this chapter discusses, packaging serves many roles to include facilitating storage and easing product transportation, as well as supporting marketing and sales.

## Product Packaging and Handling

While mapping the processes of product handling in a finished goods manufacturing support warehouse, a manager explained the reason for openings designed into the sides rather than into the tops of his product's cases. He explained that the conveyor system was fitted with suction-cup or vacuum-cup devices that would build pallet quantities of products. However, when suction was affixed to the originally designed box top opening, every so often it would pull open the top of a random box spilling the entire contents onto the conveyor line. Of course, the line would shut down for clean-up only to start again with the same event destined to reoccur. The manager was part of the team that redesigned the case openings on the sides of each box to alleviate the problem.

The example provides insight to the importance of industrial packaging for the safe and efficient handling of product through warehouse and distribution center processes. Time and space are important factors influencing the ability to cost effectively provide services to customers. Packaging can enhance or detract from this mission.

Packaging impacts product handling in several key ways.

- Facilitates efficiency through product unitizing
- Standardizes product form to allow efficient manual and mechanical movement of product
- Stabilizes products during movement and storage
- Helps to reduce materials, such as bracing and other dunnage
- Protects product, personnel, and equipment from damage during movement
- Informs of content within the packaging

## Unitization

The most basic individual unit level of measurement is a piece or an *each*. Packaging assists in the efficient handling through combining multiple pieces into single cartons, cases, or drums. Combining pieces inside a carton then allows warehouse operators to unitize individual cartons together, for example on a pallet, to create even greater economies of handling. The bundling and banding of individual pipe pieces also allows for safe and efficient product handling and storage.

## Standardization

Standardization is critical for establishing efficient routine processes and is equally important when it comes to packaging product for handling. SKUs may have various physical characteristics, but when each is packaged in a square carton, it becomes easier to manage and store. Mixed pallets containing a variety of products may achieve a level of standardization by stacking square boxes onto a pallet. Square packaging more efficiently utilizes space in slot locations within the warehouse, on pallets, and inside trailers. Handling a variety of SKUs becomes routine when packaged in cartons.

The pallet is also a form of packaging allowing for the standardization of product during movement and storage. Bulk product, such as grain or plastic-type pellets, can be poured into bags that are then set onto pallets to allow for movement and storage utilizing standard forklifts and other common machinery.

## Stabilization

Stability is important during product movement be it via a utility tractor, pallet jack, or even by hand. Packaging surrounds product so that movement is minimized between the individual items in a carton or between parts packaged as a kit. Shifting product could damage the contents and push weight to one side causing the product to tip over. Unstable packaging is hazardous in the warehouse. When visiting a Mississippi warehouse that supplied cans to a soft drink manufacturer, the manager told me of a mishap that had occurred a week earlier. He said that the warehouse was filled with empty cans ready for processing when one bay of cans shifted on the pallets causing a domino effect or chain reaction of one bay of cans falling into the other. It didn't take too long before a quarter of the warehouse was filled with thousands of cans and pallets strewn all over the warehouse floor.

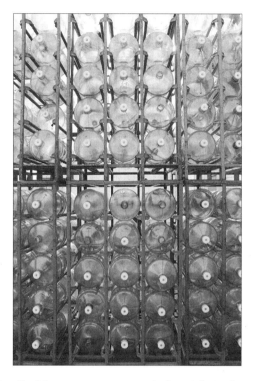

Figure 9-1    Racking system to secure industrial sized water bottles

Packaging can help to distribute product weight across a greater footprint than exists for an individual item or piece. This, too, assists in stabilizing the product. Figure 9-1 pictures industrial sized bottles of drinking water placed in steel racking to support the weight and shape of the product. While better distributed, the product's weight may, however, be too great to allow for significant stacking heights. This is particularly true for products having marketing packaging that is less ridged and sturdy. For example, although metal cans of soup help to provide stability during stacking, placing several pallets of plastic water bottles in a single stack would likely cause crushing to the bottles on the lower pallet positions.

More than 10 years ago, the manufacturer and marketer of a global baby food brand began switching from glass jars to plastic jars to help in reducing packaging and transportation costs in the supply chain. Empty jars were purchased from a supplier and warehoused until needed for production. However, with the incorporation of plastic jars, the manufacturer can produce its own jars, in-house, and significantly reduce inbound transportation costs and materials storage costs. Finished goods distribution costs were also reduced as the weight of the product was significantly less now that the weight of glass was replaced with the weight of plastic. This meant that more product could be transported without over-weighting a trailer. Fewer trailers and containers were required to ship the same amount of product to customers. This example illustrates the necessity for warehouse operators to work with marketing, manufacturing, and procurement to evaluate the cost and service trade-offs associated with packaging designs and alterations.

For warehousing, you must also be aware of how the physical environmental elements influence the integrity of packaging. Corrugated cardboard packaging absorbs moisture, and its form becomes less ridged and less stable as humidity increases. High humidity facilities may find it impossible to stack products to the maximum height of the warehouse without utilizing a steel racking system. Cost trade-offs could be calculated to assess the validity in applying stronger packaging per unit compared with investing in racks (see also Ackerman 1997, pp. 54–57).

## Efficient Packaging Materials Usage

Proper packaging can assist in the reduction of overall packaging required to stabilize and protect the load. Even automated palletizing machines, such as those depicted in Figure 9-2 are utilized to facilitate the efficient and effective stacking and tying of product cases together on pallets and slipsheets. *Slip-sheets* are heavy-ply cardboard sheets that can be cut to the size of a pallet. Dry cereal products placed in cartons can be configured onto a slip-sheet and stretch-wrapped to create a pallet-sized unit. By utilizing clamp devices on lift equipment (called a *push-pull*), product may be stacked without having to employ rigid and heavy pallets that potentially may crush cereal cartons. Slip-sheets are less expensive than wooden or reusable pallets and take up less room in the warehouse and transport equipment when shipping.

Like pallets and slip-sheets, stretch-wrap is a common packaging material utilized to add stability and security to unitized products and at an efficient cost to the warehouse, carrier, and client. Stretch-wrap is a clear plastic-type wrapping material pulled across cartons to conform and cover the product. Stretching of the wrap takes place during the wrapping application and further secures the cartons to each other composing the unit. When wrapped and palletized or slip-sheeted, the product may not require any further bracing or dunnage to keep it stable during transit. Of course, shipments destined for international transit may require the employment of additional measures to secure loads in transit. Oftentimes, specialized bags inflated with air are placed between product units to reduce the amount of empty space within the transport vehicle that could potentially cause shifting of the load. The usage of dunnage airbags when possible may help to reduce the weight and manual application of more ridged dunnage, such as wooden bracing requiring carpentry work to secure.

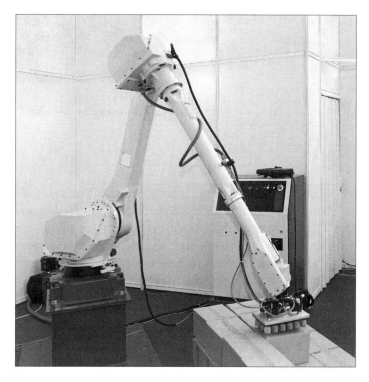

Figure 9-2    Automated palletizing machine

The new plant manager for a manufacturer of soft-top canopies for trucks recognized that the warehouse operation could reduce packaging materials, the labor required to build the packaging crates, and the cost of building the shipping crates by working with production engineering and the customer to redesign the largest structural part of the bow kit that supported the canopies of the product. A simple hinge placed in the center of the major metal arch structures would allow each piece to fold—thereby, reducing the crating materials and cubic container requirements by approximately 33percent. Figure 9-3 illustrates the new design and packaging of the equipment. This meant that the warehouse gained 33 percent more space in storing that product and could shift four of the workers who previously spent all day building crates to more productive positions. Warehouse space, handling, packaging, and labor costs were substantially reduced. In addition, the manufacturer was responsible for paying the transportation freight charges. Now, the carrier could fit nearly one-third more freight onto each flatbed trailer. Material efficiencies lead to cost efficiencies.

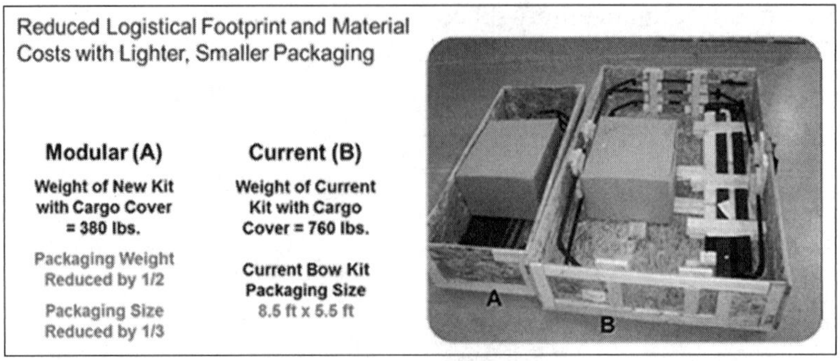

Figure 9-3    Soft-top canopy redesigned to reduce packaging and shipping costs

## Protection

Previous discussions on unitization, standardization, stabilization, and efficiency for packaging touched on safety and protection of products during warehousing and distribution operations. Certainly, one of the primary responsibilities of a warehouse operator is to protect the product entrusted to his care. Equally important is the need to protect employees of the warehouse, carriers entering the facility, and any other visitors that may be in the vicinity of stored or moving product.

Packaging may provide protection in these key forms. You need to protect product content from damage due to

- Contact friction inside a carton
- Contact with other products or obstacles outside a carton
- Shaking and dropping
- Contamination
- Pilferage and theft
- Carelessness

Dividers inside the product container provide protection so that individual items in a carton do not rub or bang together during handling. Damaged product labels and marketing packaging containing the actual product are at risk when pieces within containers lack protective packaging from each other. Similar concern exists when products meet up with other products or obstacles, such as a rack or lift device. Without protective packaging to shield contents, damage is likely.

Michigan State University possesses a machine utilized to shake or vibrate products to research thresholds of product integrity under prescribed movements. Among many research uses, the large machine could be used to study the vibration frequency at which a bracket holding a compressor within a home refrigerator would break causing the all-too-familiar rumbling noise many of us recall as a child when our family's refrigerator compressor shut off. The "shake" machine could also be utilized to test a specific packaging's capability to protect the contents of a container or package under the expected handling and transport conditions. Similar types of machines enable researchers to test the protective strength and cushioning of a product's packaging combined with the durability of the product by recording the results of a series of drops at varying heights. The heights would simulate realistic conditions under which a product may be dropped, tossed, pushed, or stopped, whether planned or inadvertent.

Packaging also serves to reduce the likelihood of one product contaminating another. Isolating a product in a carton, bag, or drum can assist warehouse operators in maintaining the identity of each SKU and keep one from physically mixing with another. Managers must also be aware that odors from one product may have the potential to permeate another product. Perhaps the container can protect permeating odors, but the typical packaging for boxed dry laundry detergent is no match for the sweet smell of chocolate. Occupying the same warehouse location or trailer, the detergent will ultimately absorb the odor of the chocolate.

Most of us probably don't enjoy thinking about pilferage and theft in the workplace, but it is a subject that is important for managers of warehousing to pay attention to. First, packaging can play a role in concealing the identity of a product. Valuable and attractive products are susceptible to theft and pilferage. Concealing the identity of 2"x 2" digital music device may help to protect it from being hidden in a pocket and walking out the warehouse door. However, it's not just expensive per unit items that are likely stolen. One warehouse had such a problem with employee "grazing," or taking edible product from a carton and consuming it during a break or lunch period, that it installed internal cameras and adopted a no exception policy on pilfering. A warehouse worker would be fired on the first incident. In another distribution facility operated by a large discount retailer, theft was so much a part of the culture of the workplace that new management installed a metal detector and an employee search policy for workers exiting the warehouse.

One theory would be that the more packaging utilized would lead to greater protection of the product. However, one high-end office furniture manufacturer and distributor found that when it utilized stretch-wrap to secure products to a pallet the employees could see the product, and this led them to exercise greater care when handling the furniture.

Managers must monitor product movement through the facility and within the control of individuals to ensure that the product is safe and employees are safe. Workers often are required to lift and move product independently and with co-workers. Packaging can assist workers if hand openings are designed into the packaging for products requiring two persons to lift. Employees will have less opportunity to come into direct contact with product if it is secured within a container. Hazards may come in the form of inhalation from airborne product or from bumping into a product that has sharp edges or corners. In addition, the structure of packaging should be designed so that product does not fall onto an employee.

## Identification and Communication

Radio frequency identification (RFID) tags provide the benefit of locating a product without requiring a line of sight to do so. This relieves worries of a scanner not reading a bar code due to it being torn or soiled. Moreover, it provides greater protection from employee errors due to misreading a product code when receiving or picking a product.

In a perfect world RFID tags would be affixed to all cartons and containers, and it would be of no consequence if a product code couldn't be identified on a case. Although more distribution center operations are utilizing RFID tags today, in reality, the industry continues to rely on a combination of RFID technology including bar codes and RFID tags, and visual identification and verification of product codes. For this reason, warehouse operators must employ a system of product identification that is user-friendly for employees assigned to handle and verify product locations within the warehouse and while entering and exiting the facility.

Package identification is likely determined by marketing, and warehousing must operate within the product codification given. In any case, product codes should be easily viewed and distinguished on all surfaces of the cartons. Most important, all hazardous identifying marks should be highly illuminated and employees well trained in identifying and handling such materials and products.

Ease of viewing and identifying product codes ensures the highest accuracy in picking and in maintaining inventory integrity. Costs, claims, and customer dissatisfaction associated with missed shipments can be reduced with greater accuracy attributed to easier product code distinctions. Identifying hazardous materials ensures employee safety and the safety of anyone needing to interact with hazardous materials and products. Regulations required specific designations for hazardous materials.

## Summary of Key Points

This chapter discussed the importance of industrial packaging to the safe and efficient handling of product through warehouse and distribution center processes. Packaging

- Assists in the reduction of overall packaging required to stabilize and protect the load
- Impacts product handling and serves to protect the product from damage

- Impacts transportation costs

- Conceals product identity to help reduce pilferage or to help prevent accidental exposure of a product to competitors

- Enhances worker safety

- Facilitates the ease of product identification and can assist in ensuring order accuracy

- Serves to support marketing and sales objectives

## Key Terms

- Distribution Costs

- Economies of Handling

- Pallet Jack

- Picking Pilferage

- Product Codes

- Protective Packaging

- Radio Frequency Identification (RFID) Tags

- SKU

- Slip-Sheets

- Stabilization

- Standardization

- Stretch-Wrap

- Unitization

## Suggested Readings

Ackerman, Kenneth B. (1997), *Practical Handbook of Warehousing*, 4th Ed., Chapman and Hall, New York, NY.

Marianne Jahre and Carl Johan Hatteland, (2004) "Packages and physical distribution: Implications for integration and standardization," *International Journal of Physical Distribution & Logistics Management*, Vol. 34 Issue: 2, pp.123–139.

Twede, D., R. H. Clarke, and J. A. Tait (2000), "Packaging Postponement: A Global Packaging Strategy," *Packaging Technology Science*, Vol. 13, No. 3: 105–115.

Murphy, P. R. and D. J. Wood, (2011), *Contemporary Logistics*, 10th ed., Chapter 11, Prentice Hall/Pearson, Upper Saddle River, NJ.

# 10

# WAREHOUSING AND TRANSPORTATION INTERFACE

## Introduction

The contribution of good communications throughout the supply chain cannot be under-estimated in terms of contribution to efficacy, efficiency, and cost reduction. A center point for critical communications and interface is the warehouse manager and operator. This individual serves as a primary interface with the driver and carrier and the customer. Not only do these interfaces directly impact operational efficiency, but they also directly affect the costs incurred and borne by various entities in the supply chain. The interface between the warehouse and the carrier and driver is so critical that often warehouses are put in charge of hiring and managing the carrier base for customers. This responsibility also requires the transportation manager within warehouse operations to be intimately familiar with the delivery and order dynamics of the customer. Regardless of where the responsibility for transportation management resides, carrier selection, contract negotiations, and maintaining a good working relationship throughout the process is key to success. Technology can assist in this endeavor from helping at the driver-warehouse worker level to ensuring that all contractual requirements are efficiently managed.

As part of the customer's supply chain, warehouses must operate in coordination with other supply chain entities. Close and frequent communication and physical interactions take place between transportation carriers and warehouses.

## Carrier to Warehouse Interaction

Many times warehouse responsibilities focus on the loading, unloading, and document exchange when interacting with carriers. Carrier selection and management are left to the customer, whereas the primary goals of the warehouse is to work with carriers to

- Set and meet scheduled delivery and pick-up appointments.

- Safely and efficiently enter, stage equipment, and exit the facility.

- Exchange proper documentation.

- Inspect freight.

- Safely and efficiently load and unload trailers.

- Minimize driver dwell time.

Driver appointments for order and load delivery and pick-up aid carriers in maximizing driving time by minimizing dwell time of the driver and equipment in the facility. Some drivers tell stories of sitting for hours only to be told at the end of a shift that they would have to layover the night and be serviced the next day. Shortsighted and solely functional-oriented warehouse managers may protect the costs to the warehouse operation at the expense of the carrier and perhaps the customer. Cost trade-offs exist between warehousing decisions and carrier decisions. Communicating with the carrier and customer may inform the warehouse operator that overtime to service the driver's load may be less costly than the carrier cost to hold a driver and equipment overnight. Moreover, a delay in the load could mean that a customer must rely on added safety stock to prevent stock-outs. The worst case, and most costly, would be for the customer to experience a production delay due to the delay of the freight, or in the case of retail and wholesale, lose the customer's business, entirely.

Progressive warehouse operations employ receiving processes that begin at the entry gate of the facility. Upon arrival, drivers may be provided a receiving door assignment from the security personnel. Warehouse and terminal information systems have the capability to utilize barcoding and RFID tag technology to immediately document the arrival of the carrier, the specific freight, and orders within the carrier's equipment, and notify receiving personnel of the arrival. Receiving doors may be assigned based on a number of factors including available doors unoccupied, location of putaway locations for the freight after unloaded, and even the need for cross-docking a portion of the freight assigned to an outbound order and staging area. Again, the warehouse operator must interact with carriers to accomplish the process and cost goals of the warehouse, carrier, and the supply chain as a whole.

After entering the facility, drivers must interact with receiving personnel to verify documentation and inspect freight. The driver's delivery receipt cannot be signed by the receiving clerk and processed if the documentation, bill-of-lading, and manifest fails to match perfectly the physical freight being delivered. Supply chain and customer-oriented warehouse managers work closely with drivers, carrier managers, and the client and customer of the warehouse to resolve discrepancies. Doing so helps to reduce costs, such as

- Carrier costs including driver and equipment delay, returns, and redelivery.

- Warehouse costs including gate and receiving personnel sunk-costs associated with the first delivery attempt and rescheduling appointment.

- Customer costs including reverse logistics costs to return freight and determine disposition. Redelivery costs passed on to the customer by the carrier and warehouse and costs associated with delaying the product's availability.

Left only to the driver to resolve, total supply chain costs go up and service effectiveness goes down.

**Figure 10-1    Driver interaction with warehouse personnel affects costs, safety, and schedule**

Loading and unloading processes most often require warehouse personnel. Figure 10-1 illustrates the unloading of a soft-side trailer by a warehouse terminal forklift operator. Floor loads are hand-loaded or unloaded with the assistance of a conveyor drawn into a trailer, whereas palletized or unitized loads require forklift operators. Again, nonproductive carrier time includes down time when the freight and transportation equipment is not in motion. Warehouse operators influence the driver's ability to achieve maximum productive driving time.

Drivers are responsible for the weight distribution within the transport equipment and adhering to maximum legal weight restrictions for the specific equipment. Loaders must be well trained to ensure load stability and protection of the driver and freight during transit. Many times trailers are loaded without driver supervision. When loaded the carrier is notified and a pick up time is authorized. A driver may be forced to redistribute her load if weigh scales indicate too much weight bearing down on a single axel. Drivers never want to hear that they have too much weight and must remove enough to be legal.

# Managing the Carrier Base

Some warehouses are put in charge of hiring and managing the carrier base for customers. Entire traffic departments may be employed by a warehouse operator. The interactions between warehouse and carrier previously discussed are also applicable to a warehouse responsible for transportation management. However, many more responsibilities are taken on by the warehouse operator when hiring and managing transportation for clients. The following list contains some of the added critical responsibilities.

- Carrier selection and contract negotiation
- Negotiating freight charges with client
- Documentation creation
- Ensuring carrier insurance and liability
- Auditing freight bills and initiating freight payment
- Claims management

Decisions to hire carriers should be placed in the hands of managers experienced in evaluating transportation costs and services, drafting contracts, and negotiating with carriers. Prior to signing contracts to hire carriers, the transportation or traffic manager employed by the warehouse must fully evaluate the delivering and ordering dynamics of the customer.

For example, consider the shipping of full truckloads of a single SKU containing full pallets. Loading and unloading requires a forklift, whereas counting is relatively easy if the warehouse operator knows the tie-high of the SKU full pallets. The *tie* is the configuration of a layer of product on the pallet that when every other layer is pivoted assists interlocking cartons together on a pallet. Common ties and the specific product counts for each is easily recognized by industry personnel, and when multiplied by the number of product layers on the pallet, the total product count for the pallet is determined. This makes checking the product count easy. In addition, a full truck load (TL) originating at the warehouse and destined for the customer has no intermediate cross-docking needs similar to that required of less-than-truckload (LTL) orders. A TL order such as this would require minimal handling and would allow the carrier to fully utilize the capacity of the transport equipment and maximize the driver's productive time in transit. Consequently, the carrier selected would be a truckload carrier with freight rates lower than the rates charged per unit for moving an LTL order.

Managers must evaluate the type of freight to determine the appropriate mode of transportation to utilize either motor carrier, air, water, rail, or even pipeline. Intermodal shipments are transported utilizing more than one mode for transport. The carrier may achieve low costs compared to the competition by employing an intermodal strategy;

whereby, the shipment may be picked up at the warehouse via the carrier's trucking division, transported via the carrier's rail services for the linehaul or longest leg of the transit, and finally delivered to the customer's business by motor carrier. U.P.S. is known to utilize rail intermodal services to transport some of its UPS trailers via flatcar from region to region. Trailer on flat car (TOFC) is an intermodal movement utilizing the flexibility of motor carrier and the fuel efficiency of rail to reduce transportation costs and manage consistent service for customers.

## Pricing and Negotiations

With the 1970s and '80s reduction of regulation governing pricing and service offerings allowed by all modes of carriage, traffic managers are equipped with a host of options for selecting modes and negotiating favorable transportation rates. A nonasset-based freight broker is many times a viable option by offering the best freight rate for a shipment by negotiating with the carrier on behalf of the warehouse. For example, C.H. Robinson (CHR) is a leader in facilitating shipper and carrier exchanges. CHR may help a steel hauler find a load out of Mobile, while at the same time help an importer obtain the best shipping rates for moving a load of steel from the Port of Mobile to a local steel mill. CHR also provides much more than single-load shipment negotiations. It maintains relationships with a large customer base and carrier base, and can work with each to create corporate-wide supply chain logistics solutions.

Traffic managers for the warehouse collect information about potential carriers and the services offered through the Request for Information (RFI) process. Operating authority, ensuring insurance, equipment and capacity, and lanes serviced are all important to know before moving a carrier to the next stage in the selection process. Upon selecting a pool of candidates to proceed, the traffic manager initiates a Request for Proposal (RFP). Greater description of products to be transported, volumes, value, frequency, and other services required and expected are provided to the candidate carriers. The carriers then provide greater description of their strategies to service the freight and provide detailed documentation, references, and service quality documentation as support for their application. Again, the pool is reduced to a final set of candidates that will then be asked to provide a final price quote via a Request for Quote (RFQ) for managing shipments.

Initial stages of the selection process may include all potential carriers in a collective meeting to efficiently distribute information about the warehouse transportation needs. Carriers may be asked to visit the warehouse to receive first-hand information from various managers within the warehouse and to see the product to be transported. When the final selection of candidates is determined, the traffic manager may enter into one-on-one meetings with individual carriers to finalize negotiations of service and pricing.

Although the freight-bidding process, as it is sometimes termed, may be managed a little differently from one shipper or warehouse to another, it is imperative that shippers and carriers understand the product, service, and market dynamics so that neither enters into an agreement that they cannot fulfill. Intermodal carrier strategies may not provide the time-sensitive transit required by a shipper operating on a just-in-time replenishment strategy. Efficient and low-cost carrier strategies to utilize intermodal rail services with a guaranteed 3-day transit time must be targeted toward the customer market segment that finds a 3-day delivery cycle appealing and acceptable.

It is critical for the traffic manager to understand the factors influencing the carrier's rate quotes. Product characteristics, such as, weight, cube, value, susceptibility to damage or theft, and distance from origin to destination all play critical roles in determining the cost to move the freight and must be covered by the freight rate. Market factors that also influence the carrier's rate quote include the level of volume the carrier already has within specific lanes, the fuel prices in certain regions, the availability of backhauls or returning loads, and the level of competition between carriers and among modes of transportation within each lane. The better the traffic manager understands transportation pricing from the carrier's perspective, the better he can be in the position to negotiate the best price and service for his warehouse and clients.

Motor carrier pricing flexibility came about with the passage of the Motor Carrier Act of 1980 giving carriers rate-making freedom that was highly regulated prior to the Act. Carriers could now negotiate volume discounted rates based on the shipper's freight tonnage available to the carrier. Origin to destination, or point-to-point TL rates can be established between carriers and shippers.

LTL Carriers and shippers find the National Motor Freight Classification (NMFC) system effective for establishing a common understanding of the freight description and shipping requirements to manage the freight. Product groupings are determined by the density of a product, the ease or difficulty of handling a product, the stowability or ease by which a product may efficiently be stored inside a trailer, and the liability/value of the freight. Instead of having to negotiate rates based on every type of product in existence, carriers and shippers may agree on a freight classification for the products shipped. Products with similar characteristics in the four primary factors can be thought of as equal when transporting. Product descriptions are grouped from freight classification 50 to 500 and may be placed within one of the NMFC categories for evaluating freight rating. The higher the freight classification, the higher the associated freight rate.

Although the freight classification helps to understand the product characteristics, the tariff contains the actual shipping rates based on the determined freight classification and is specific to a freight lane from origin to destination and the weight of the shipment. Freight rates may be negotiated based on limiting the carrier's liability. By lowering the declared value, a lower rate may be negotiated. This limits the carrier's financial risk in case the shipment is lost or damaged; only the lower declared value can be claimed against the carrier. A higher declared value draws a higher freight classification and freight rate. Freight with a highly varying weight to cube ration (density) can likely demand a higher price for transporting. For example, packaged inflated soccer balls would be inefficient in utilizing the capabilities of the transportation equipment. Shipping uninflated balls stacked in boxes on pallets would be more efficient for the carrier and shipper, and the classification would certainly be lower compared with the inflated balls. One shipper nested chairs together to reduce the cubic dimensions for a single-boxed chair. The cubic space was cut in half because the chairs could be nested, and the density of the "new" single box (containing two chairs) received a lower freight classification.

The point of this discussion is to encourage traffic managers within the warehouse to understand the product and market characteristics that carriers utilize to establish freight rates. Doing so helps both parties to reach successful agreements for managing the customers' transportation needs. Failing to do so causes the traffic manager to be at a significant disadvantage when negotiating rates with carriers.

Other influencing factors of transportation pricing include regular and guaranteed high-volume transport of a single commodity between two points: commodity rate. Along with limiting the declared value, reducing the driver's responsibilities for loading and unloading can significantly reduce transportation rates. In consumable packaged good distribution, the term freight-all-kinds (FAK) is used to classify different freight types but that have general transport similarities. Although they may be classified in a slightly higher or lower freight class, the shipping volume per item does not warrant separate classifications. The group of products may be placed in a FAK, for example, class 55.

Surcharges, demurrage and detention, and accessorial charges all have the effect of raising the overall final freight bill to be paid. Although in most cases it is likely that a shipper would want to limit such additional charges. However, it is common practice for shippers to utilize chemical railcars as temporary rolling warehouses. Demurrage charges can be much less costly than the investment required to expand permanent chemical storage facilities.

# Bill of Lading

Traffic departments within the warehouse may also be responsible for drafting and printing Bills of Lading (BOL) and shipment manifests for outbound loads. They must also provide a delivery receipt or signatures for inbound loads. Figure 10-2 provides an illustration of a BOL. BOLs govern the shipment during transit and delivery from origin to destination. Critical shipment information is contained on the legal document and pertains to the shipper (consignor), receiver (consignee) and carrier. Freight class and description are contained on the BOL. Weight, cube, piece count, billing information, special instructions, declared value, COD charges, and contact numbers are contained on the BOL. Shippers and carriers sign the BOL to indicate acceptance of the freight in proper condition and possession of the freight. The BOL serves as a document delivery receipt, proof of ownership, and title of goods transfer document and is used to initiate freight bills, payment, and claims.

| Carrier | Company X | |
|---|---|---|
| Shipper<br>Address:<br>City:<br>State / Zip:<br>Phone: | Pick Up Date & Time | Bill Third Party To: |
| Consignee<br>Address:<br>City:<br>State / Zip:<br>Phone: | Delivery Date & Time | Bill Charges To: |

Special Instructions

| # of Packages | Pkg. Des. | *HM | Description | NMCF Item# | Class | Weight (lbs) |
|---|---|---|---|---|---|---|

| P.O. Number (other references if applicable) | Total Weight ><br>Subject to correction | XX (lbs) |
|---|---|---|

Hazardous Materials Contact Number:

| C<br>O<br>D | COD Amount<br><br>U.S.$ | Remit To:<br><br>Note - When the rate is dependent on value, shippers are required to specify in writing the agreed or declared value of the property. The agreed or declared value of the property is hereby specifically stated by the shippers not to exceed. |
|---|---|---|

Received at the point of origin on the date specified, from the consignor mentioned herein, the property herein described, in apparent good order, except as noted (contents and condition of contents of packages unknown), marked, consigned, and destined, as indicated above, which the carder agrees to carry and to deliver to the consignee at the said destination, if on its route or otherwise to deliver to another carrier on the route to said destination. It is mutually agreed as to each carrier of all or any of the goods over all or any portion of the route to destination, and as to each party of any time interested in all or any of the goods, that every service to be performed here under shall be subject to all the conditions not prohibited by law, whether printed or written, are hereby agreed by the consignor and accepted for himself and his assigns.

I hereby declare that the contents of this consignment are fully accurately described above by proper shipping name and are classified, packed, marked and labeled, and are in all respects in proper condition for transport by rail, water according to applicable international and national government regulations.

| SHIPPER | CARRIER |
|---|---|

Figure 10-2  Example of a Bill of Lading

# Freight Payment and Claims Management

Managing freight payment of hundreds of daily orders and shipments can be a cumbersome job. Warehouse management systems (WMS) have helped make the job more manageable because some may enable the comparison of the freight bill to the agreed upon charges. This may require correlating freight bill information with critical information and notations on the original order, BOL, and delivery receipt. The task can be more difficult if items were substituted or lost or over-shipping occurred. Moreover, add on accessorial charges, detention, and any additional surcharges, and the auditing process can be overwhelming.

Freight payment services have been around for decades and provide experienced freight bill auditing and payment. Cass Information Systems, Inc., for example, provides freight bill auditing, payment, claims management, and a suite of additional logistics and transportation services for warehouses, shippers, and carriers. Restoration Hardware, Pepsi, Dole, and Chiquita utilize Cass services. Third-party freight payment and other information-based services filled an industry need because warehouses, manufacturers, and marketers had expertise in managing freight, production, and selling products and services but were not necessarily experts in managing freight auditing and payment functions.

As consumer and industrial demand increases both domestically and worldwide, demand for transportation services and the complexity and creativity in shipping and routing will exponentially increase. Cash-to-cash cycles can improve as freight bills are paid at the optimal time; not too early to miss-out on opportunity cost of capital but not too late to miss-out on payment discounts or to incur late payment fees. Traffic managers for the warehouse may be in the position to determine if freight bill auditing and payment should be performed in-house or outsourced. The correct decision is the decision that renders better cash management resulting in higher levels of working capital.

## Traditional Cost Trade-Offs Between Warehousing and Transportation (This Section Is Based on Murphy and Wood, 2011, pp. 133–146)

Many warehouse operators are not in charge of managing client or company inventory policies and procedures pertaining to how much and when to order, or how much stock to warehouse. Others play a more direct role in the decisions influencing inventory inbound and as it relates to outbound orders. In either case, warehouse managers will be better informed and make better decisions that impact internal and external inventory levels. This section discusses the role of inventory as it pertains to the interaction of the warehouse and carrier.

The classic simple economic order quantity (EOQ) model establishes the order quantity to place with a supplier that is associated with the lowest total cost when considering only the cost to order and the cost to carry the inventory. EOQ is cycle stock or the amount to order each cycle that will satisfy known demand. As long as the railcar, motor carrier, air freighter, or barge arrives on time, the EOQ will be the inventory required each order period.

Several grand assumptions exist that make the EOQ a good starting point for assessing inventory needs; but adjustments need to be made when customer demand changes, transportation or production lead-times fluctuate, and large volume discounts are offered by vendors and carriers. Carrier reliability to pick up and deliver on time is such an important service quality factor, so it is a point in which warehouse operators and carriers must understand the impact reliability has on inventory levels.

For example, assume that ordering in a product lot size of 200 units (EOQ) can render the lowest total cost when combining the cost to order each time and the inventory carrying cost associated with the EOQ. Further suppose that a safety stock level is set at 75 units to protect against a stockout if the carrier fails to arrive on the required delivery date. Utilizing a fixed order quantity procedure, assuming that average daily sales is 25 units and the average order cycle time requires 3 days to receive an order from the time it has been placed, the reorder point will be when the stock level on hand lowers to 150 units.

> Reorder point: 150 units = (daily sales 25) (transit time 3 days) + (75 units safety stock)

When inventory on-hand reaches 150 units, an order is placed for the EOQ (200 units). This is in anticipation of the truck arriving in 3 days as expected. If all goes as planned, the truck will arrive just as the final unit of cycle stock is sold. However, if the carrier is late, the warehouse operator must dip into safety stock to meet sales demand during the time the truck is late. Higher levels of safety stock are necessary when the carrier's reliability is poor and steadily declining.

The traffic manager and the warehouse manager must work together to resolve carrier issues. A late delivery or missed appointment may be due to the carrier stopping first at other locations to deliver prior to delivering to the warehouse. The carrier may have had difficulty getting unloaded in a timely manner at the warehouse in the past, so the carrier began scheduling the warehouse as the last stop in the delivery cycle. Fluctuations in delivery are then influenced by the time required to unload at all the other stops. Some days the final delivery cannot be made on time. However, perhaps it is a carrier that is lacking appropriate equipment capacity to perform the job on time. In either case, the traffic manager, warehouse manager, and carrier must pull together to resolve the root causes of the carrier's delay. Otherwise, higher levels of safety stock will be required to protect against stockouts. Consequently, inventory costs will increase without an increase in sales. Inventory turns will suffer and so too will return on assets.

## Varying Roles of Warehouse and Carrier

It is clear that to be competitive and add value for customers and the firm, warehouses and carriers must work together to achieve success and reduce conflict. Throughout history, shippers (in this case the warehouse) have pushed for the lowest transportation rates, whereas carriers have pushed for rate increases. Operating costs for carriers are on the rise, and if operating revenues are suppressed, the operating ratios for even the strongest carriers will be in jeopardy.

Operating ratio is one measure of the carriers' capability to manage cost and revenue to achieve operating efficiencies. It is calculated as (Operating expenses / Operating revenue) x (100). Knight Transportation's shareholder's report, for example, indicates the company's 2012 operating ratio is 88.4 percent (Knighttrans.com). Therefore, of every dollar of revenue generated through freight transportation services (excluding nontransportation income), Knight Transportation retains 11.6 cents that can go to pay interest and returns to stockholders. In an industry in which average operating ratios fall within the mid-90s, Knight Transportation proves to be an efficient motor carrier.

**Responsible Decision Making**

At Knight Transportation, we know our decisions affect the world we all live in. Our promise of Delivering More includes our significant investments in technology that eliminate waste, promote efficiency, and reduce our carbon footprint. Knight's commitment to offering sustainable solutions is real. These efforts include

- Fuel-efficient tractors
- Fuel conservation initiatives
- Lower tractor emissions
- Energy-saving solar farms
- Alternative fuel R&D
- Aerodynamic trailer technology

When you choose to do business with Knight, you can feel confident your dollars are being reinvested into sustainable technologies. Efficiency = Sustainability. (Source: www.knighttrans.com/Sustainability/index.html)

The point is that efficient carriers focused on sustainability should be good carrier partners. Carriers with higher operating ratios need to seek increased freight rates or work with shippers and warehouses to find solutions that help to improve efficiencies and lower operating expenses. Because of the connectivity of transportation to the warehouse business, both entities must work together to achieve a sustainable future.

## Shipper Load and Count (SLC)

One way warehouses can assist carriers in reducing operating expenses is to help reduce the time they must spend at the warehouse picking up and delivering. Shippers that have "partnership" type relationships with carriers may find it beneficial without significant risk to allow the warehouse to load trailers without requiring carriers to count the freight. The shipper then assumes responsibility for the count: shipper load and count (SLC). This doesn't relieve carriers from maintaining the freight within their possession. They still have a legal obligation to protect the freight. On the receiving end, the consignee could also relieve the carrier from counting the freight and expedite the unloading process. Full truckload quantities transported to a single destination would be the best setup for an SLC program.

## Terms of Sale/Transportation[1]

Terms of sale stipulated on the purchase order and in the contract between vendor and customer designate the party in control of and responsible for setting up and managing the carrier for the shipment. Moreover, the terms of sale indicate when ownership changes hands from seller to buyer and stipulates who pays the carrier's freight bill.

It may seem at first that shipping would be made simpler if the customer allowed the shipper to be responsible for the shipment, pay the carrier, and deliver the freight to the destination before the customer took ownership and responsibility for the product. However, there are many reasons that one or the other party would want to control the shipping terms of sale:

- The party has additional freight to add to the shipment and thus reduce the rates for the shipment.

- The party has well-established relationships with quality carriers and competitive rates.

- The party wants to sell the product and combined transportation services in one single charge as a revenue-generating service for the receiver.

There are also reasons against controlling the shipment in transit:

- The controlling party assumes liability risk and owns inventory in transit and costs associated.

- The controlling party files and manages claims for damages and shortages in transit.

- The controlling party must manage the loading, carrier, and administrative costs associated.

Terms of sale can be an equitable negotiating factor for either side depending on the expertise in managing transportation services. Table 10.1 illustrates the common Free on Board (FOB) domestic transportation terms of sale.

Table 10-1    Domestic FOB Terms of Sale

|  | FOB Origin Freight Collect | FOB Origin Freight Prepaid | FOB Origin Freight Prepaid and Charged Back | FOB Destination Freight Collect | FOB Destination Freight Prepaid | FOB Destination Freight Prepaid and Charged Back |
|---|---|---|---|---|---|---|
| Owner in Transit | Buyer | Buyer | Buyer | Seller | Seller | Seller |
| Pays Freight | Buyer | Seller | Seller pays and bills buyer | Buyer pays on arrival | Seller | Seller pays and bills buyer |

# International Transportation

Table 10-2 illustrates the International Chamber of Commerce, 2010 INCOTERMS (International Commercial Terms) governing the international terms of sale. Similar to the intent of domestic terms of sale, INCOTERMS serve to assist carriers, shippers, consignees, financial institutions, and others with interest in international freight transportation in "speaking in the same logistical terms." It reduces issues stemming from disagreements in ownership, responsibility, and liability for freight traveling over international boundaries, which compared with domestic freight movement can be daunting to accomplish (see Figure 10-4).

**Figure 10-4   Yangshan Deep-Water Port-Shanghai, China: World's Largest Container Port System**

Complexity is magnified by the amount of documentation required to transport freight between countries having differing laws and documentation requirements. Table 10-3 highlights the core documentation required to facilitate successful international freight transportation. Globally focused logistics service provider Crane Worldwide Logistics (www.craneww.com) partners with customers to provide "Full-Service Air, Ocean, Trucking, Customs Brokerage and Logistics" services anywhere in the world. Expertise in international documentation and freight handling is a must for controlling costs and meeting service expectations.

**Table 10.2    INCOTERMS 2010 (Derived from International Chamber of Commerce,**
**www.iccwbo.org/products-and-services/trade-facilitation/incoterms-2010/)**

| INCOTERMS 2010 (eleven terms) | Example | Key Aspect |
|---|---|---|
| EXW – ExWorks | EXW Guangzhou City, China (City of Plant pick-up) | Transfer at PLANT. Buyer loads on truck. Buyer bears cost and risk. Minimal obligation to seller. Used for all modes. |
| FCA – Free Carrier | FCA Waterford, Ireland (City of Truck Terminal) | Deliver to TRUCK/Co. in Waterford. Buyer selects carrier. Seller loads on TRUCK or TERMINAL. Seller clears for export. Used for all modes. |
| FAS – Free Alongside Ship | FAS Yangshan, China (City of Loading Port) | Deliver to PORT & unload. Buyer assumes cost and risk from that point. Buyer clears for export. Used for sea or inland waterway. |
| FOB – Free on Board | FOB Mobile, USA (City of Loading Port) Variant: "stowed and lashed" | Delivered over SHIP's rail. Buyer assumes cost and risk from that point. Seller clears for export. *GRAY AREA: Cost to stow/lash* Used for sea or inland waterway. |
| CFR – Cost and Freight | CFR Kaohsiung, Taiwan (City of Discharging Port) Variant: "landed" meaning cost to discharge. | Seller pays cost and freight charges to transport to named port of DISCHARGE DESTINATION. Buyer assumes risk of loss when delivered over SHIP's rail-port of shipment. Seller clears for export. Stowed and lashed. Insurance paid by Importer. *GRAY AREA: Cost to discharge.* Used for sea or inland waterway. |

| INCOTERMS 2010 (eleven terms) | Example | Key Aspect |
|---|---|---|
| CIF – Cost, Insurance & Freight | CIF Long Beach, USA (City of Discharging Port) Variant: "maximum coverage, landed" if want above CIF amount. | SAME as CFR. PLUS SELLER PAYS INSURANCE until Dest. Port. Base insurance of 110% cargo value GRAY AREA: Cost to discharge. Used for sea or inland waterway. |
| CPT – Carriage Paid To | CPT Detroit, USA (City of Destination) | Multimodal (intermodal) Exporter Trans. cost to Dest. City. Risk transfers to buyer when loaded onto carrier. Exporter selects carrier. Seller clears for export. Not unloaded. Used for all modes. |
| CIP – Carriage & Insurance Paid to | CIP Detroit, USA (City of Destination) Variant, Variant: "maximum coverage, landed" if want above CIP amount. | SAME as CPT. PLUS seller provides INSURANCE until Dest. City. Base insurance of 110% cargo value. Used for all modes. |
| DDP – Delivered Duty Paid | DDP San Diego, USA (Final Destination City) | All modes of transportation. Exporter risk and trans. cost to Final Dest. City and Facility. Exporter selects carrier. YES exporter clears import Customs. YES exporter pays duty. NOT Unloaded. Could use a variant, "Unloaded." |

| INCOTERMS 2010 (eleven terms) | Example | Key Aspect |
|---|---|---|
| DAT – Delivered At Terminal | DAT Pensacola, USA, Warehouse 1. (Named terminal at the named port) | All modes of transportation. Exporter risk and transportation cost to named terminal and port. Specify in contract exactly where in the terminal risk and cost changes hands. *YES* Unloaded. Exporter clears for export. Importer clears for import. Importer pays import duty. Note: If parties agree for Exporter to clear import customs and pay import duties, then DDP (Delivered Duty Paid) may be the best INCOTERM. |
| DAP – Delivered At PLACE | DAP Atlanta, USA, IKEA Import Warehouse. (Named place.) | All modes of transportation. Exporter risk and transportation cost to named place of destination ready for unloading. *NOT* Unloaded. Could use a variant, "Unloaded." Specify in contract exactly where at the place the risk and cost changes hands. Exporter clears for export. Importer clears for import. Importer pays import duty. Note: If parties agree for Exporter to clear import customs and pay import duties, then DDP (Delivered Duty Paid) may be the best INCOTERM. |

Table 10-3  International Commercial Documentation (Based on David and Stewart 2010)

| Invoices | Use | Specifics of Document |
| --- | --- | --- |
| **Commercial Invoice** | For importer direct or through bank. | Contains what is being billed. |
| | | Product description. |
| | | Harmonized System #. |
| | | # of units. |
| | | Dimensions. |
| | | Weight. |
| | | Total value. |
| | | Incoterm(s). |
| | | List of prepaid items. |
| | | Terms of payment. |
| | | Accompanied by LC. |
| | | Currency of payment. |
| | | Shipping information. |
| | | Ports. |
| | | Companies. |
| | | Dates. |
| | | Number of units. |
| | | Weight (gross & net). |
| | | Size of shipment. |
| | | Seller-exporter. |
| | | Buyer-importer. |
| | | Contact name & address. |
| **Pro-forma Invoice** | For importer and "issuing bank" to issue LC. | Not an invoice, but a quote. |
| | | Must be accurate and precise. |
| | | Used for quote comparing. |
| | | Used for obtaining LC. |
| | | Same info as on commercial invoice. |
| | | Must include expiration date of "quote." |
| | | Quote cannot be withdrawn before expiration date. |
| | | Can be accepted any time until expiration date. |
| **Consular Invoice** | For the import country's Consulate. | Same as commercial invoice, but printed on import country's Consulate stationary and visa-stamped. |
| | | Obtained before being sent to importer. |

| Invoices | Use | Specifics of Document |
|---|---|---|
| **Specialized Invoice** | For import country's Customs Office. | Same as commercial invoice, but printed on specific standard international form to standardize for Customs. |

| Export Documents | Use | Specifics of Document |
|---|---|---|
| **Export License** | For the export country's government. | Authorization to export a specific product. Controls national treasures and foreign trade for military or political purpose. |
| **Shipper's Export Declaration** | For U.S. Customs Service pertaining to U.S. exports. | Exports valued over $2,500 per item category (or $500 for parcels sent via postal. Determined by Harmonized System Number. Data gathering to determine what and where sold. Goes into National Trade Data Bank (www.stat-usa.gov). Must be with U.S. Customs before shipment. Primarily submitted through the Automated Export System. Here the "Exporter" is always the "manufacturer," but, it is the importer's responsibility to file. |
| **Certificate of End-Use** | Provided by the import government to the export government. | Certifies the product will be used for legitimate purposes, say military training. Certifies product will not be diverted and used for unacceptable purposes such as for civil law-keeping. |

| Import Documents | Use | Specifics of Document |
|---|---|---|
| **Certificate of Origin** | Used by import Customs. | Does not stipulate where manufactured, but only where shipped from. Statement that goods originated (shipped) from a certain country and signed by the export Chamber of Commerce. Used to assess tariffs, quotas, and compile trade statistics. Different countries pay different tariffs, so it may be misused if indicated shipped from place other than where manufactured. |

| Invoices | Use | Specifics of Document |
|---|---|---|
| **Certificate of Manufacture** | Also, used by import Customs. | Similar to the Certificate of Origin, but specific to origin of manufacture. |
| | | Statement that goods were manufactured in a certain country and signed by export Chamber of Commerce. |
| | | Used to assess tariffs, quotas, and compile trade statistics. |
| **Certificate of Inspection** | Requested and used by importer. | Created by a third party that has inspected goods content, quality/condition, and quantity, and provided a preshipment inspection. |
| | | Evidence that there is no problem with the product. |
| | | SGS company from Switzerland is the largest inspection company. |
| | | Important to have for Document Collection and LC. |
| | | Inspection can include determining correct value of product for invoicing and tariffs. |
| **Certificate of Certification** | Required by import country. | Sometimes called Certificate of Conformity. |
| | | Certifies that product can pass certification procedures. |
| | | Produced by third party (sometimes by exporter and signed by Chamber of Commerce). |
| | | Used to specify that product meets the required technical standards of the importing country. |
| **Phyto-Sanitary Certificate** | Required by import country. | Produced by third party or Agriculture & Food Safety agency of export country. |
| | | Used to ensure product is free of disease and is not infested. |
| **Certificate of Analysis** | Requested and used by importer. | Produced by independent laboratory. |
| | | Used to ensure accurate composition of mixtures, such as, chemicals, cement, alloys, and polymers. |
| **Certificate of Free Sale** | Written and signed by exporter and export Chamber of Commerce (or government agency regulating the product). | Certifies that it is legal to sell the product in the country of export. |
| | | Reduces chance of exporter selling inferior product that could not be sold in the export country. |
| | | Common in pharmaceuticals. |

| Invoices | Use | Specifics of Document |
|---|---|---|
| **Import License** | Required by government of developing countries and provided by importer. | Authorization to import the product. Developing countries may want to keep out luxury goods when they have a short supply of foreign currency, and spend such currency on goods to help its economic position. |
| **Consular Invoice** | For the import country's Consulate. | Same as commercial invoice but printed on import country's Consulate stationary and visa-stamped. Obtained before being sent to importer. |
| **Certificate of insurance** | Depends on Incoterm. Can be required by importer or import country. | Obtained from the insurance company covering the product. |

| Transportation Docs. | Use | Specifics of Document |
|---|---|---|
| **Ocean Bill of Lading** | Drafted by carrier for shipper, carrier, and consignee use. | Contract of carriage. Used for containers, autos, crates, and products not commissioning the entire ship. Contract of carriage: between shipper and carrier. Receipt for goods. Certificate of title. Straight BL: has defined consignee. To order BL: designated with "consigned to order" in the consignee blank and is negotiable as to who is ultimate owner of goods. For product sold during shipping. |
| **Uniform Bill of Lading** | Drafted by carrier for shipper, carrier, and consignee use. | Used for inland transportation. |
| **Intermodal Bill of Lading** | Drafted by carrier for shipper, carrier, and consignee use. | Recent document used for intermodal moves covering multiple carriers. |
| **Air Waybill** | Drafted by carrier for shipper, carrier, and consignee use. | Specific to airfreight. Straight BLs. |

| Invoices | Use | Specifics of Document |
|---|---|---|
| **Charter Party** | | In place of the Ocean BL for bulk commodities, such as, oil, ores, grain, polymers, sand, cement, sugar). |
| | | Negotiated for: voyage charter or time charter. |
| | | Negotiated for bareboat charter. |
| **Packing List** | Provided by exporter and accompanies shipment. | Identifies precise number of containers and precise SKUs and product description in each. |
| | | Helps to avoid inspection delays. |
| | | Commercial invoice can substitute for this. |
| **Shipper's letter of Instruction** | Provided by shipper to the carrier. | Contains special instructions for cargo handling during transport. |
| **Shipments of Dangerous Goods** | Generally prepared by the shipper as a "Shipper's Declaration of Dangerous Goods." | Specific regulations over dangerous goods. |
| | | Best prepared by expert. |
| | | Also noted on BLs. |
| | | Implications for packaging, labeling, and stowing. |
| **Manifest** | Prepared by the carrier for internal use, but can be requested by government authorities. | List includes all cargo, ownership, port of embarkation and debarkation, and special instructions. |

# Summary of Key Points

As with any dynamic process containing numerous moving parts, good communications up and down the supply chain is key to success. As this chapter shows, there is no more critical interface in the supply chain than at the warehouse-carrier and driver point. The warehouse manager must ensure this interface is optimized using soft (people) skills and the application of available technology. The interface between the warehouse and the carrier and driver is so critical that often warehouses are put in charge of hiring and managing the carrier base for customers. Regardless of where this responsibility resides, the responsible manager must be a seasoned one with experience due to the nature of duties involved. From determining the requirements and selecting the carrier to managing the contract and the communications flow, this is a challenging task, for instance:

- It is critical for the traffic manager to understand the factors influencing the carrier's rate quotes.

- Managers must evaluate the type of freight to determine the appropriate mode of transportation to utilize.

- The better the traffic manager understands transportation pricing from the carrier's perspective, the better he can be in the position to negotiate the best price and service for his warehouse and clients.

- The importance of carrier selection and the establishment and maintenance of a good relationship and open communications channel cannot be overemphasized for the contribution to operational success.

- International shipments are more complex and require greater expertise in international documentation and carrier management. Expert logistics service providers such as Crane Worldwide Logistics can help customers better manage the international terms of trade and commercial documentation.

## Key Terms

- 1980 Motor Carrier Act
- Accessorial Charge
- Backhaul
- Bill of Lading (BOL)
- Cash to Cash Cycle
- Commodity Rate
- Contract
- Cross-docking
- Cycle Stock
- Demurrage/Detention
- Dwell Time
- Economic Order Quantity (EOQ)
- Fixed Order Quantity
- Floor Loads
- Free on Board (FOB)
- Freight Bidding Process
- Freight Bill

- Freight Bill Auditing
- Freight Payment
- Freight Rate
- Freight-All-Kinds (FAK)
- Full Truck Load (TL)
- INCOTERMS
- Intermodal
- Inventory Carrying Cost
- Inventory Turns
- Just-in-Time Replenishment
- Lead-time
- Less Than Truck Load (LTL)
- Linehaul
- Manifest
- Motor Carrier
- Operating ratio
- National Motor Freight Classification (NMFC)
- Nonasset Based Freight Broker
- Outbound Order
- Outsource
- Putaway
- Radio Frequency Identification (RFID) Tags/Technology
- Receiving Clerk
- Request for Information (RFI)
- Request for Proposal (RFP)
- Request for Quote (RFQ)
- Reverse Logistics
- Safety Stock
- SKU

- Stockout
- Sunk Cost
- Supply Chain
- Surcharge
- Tariff
- Tie-High
- Trailer on Flat Car (TOFC)
- Transportation/Traffic Manager
- Unitized Load

## Suggested Readings

Closs, D. J., S. B. Keller, and D. A. Mollenkopf. (2003), "Chemical Rail Transport: The Benefits of Reliability," *Transportation Journal*, V. 42, No. 3: pp. 17–30.

Coyle, J. J., C. J. Langley, , R. A. Novack, and B. J. Gibson. (2013), *Supply Chain Management: A Logistics Perspective*, 9th ed., Chapter 10, South-Western/Cengage Learning, Mason, OH.

David, P. A. and R. D. Stewart. (2010), *International Logistics: The Management of International Trade Operations*, 3rd ed., Chapter 9, Cengage Learning, Mason, OH.

Hazen, J. K. and C. F. Lynch, *The Role of Transportation in the Supply Chain*, CFL Publishing, Memphis, TN.

International Chamber of Commerce: The World Business Organization, www.iccwbo.org/products-and-services/trade-facilitation/incoterms-2010/.

Lynch, C. F. (2004), *Logistics Outsourcing*, 2nd ed., CFL Publishing, Memphis, TN.

Notice of Annual Meeting to be held on May 16, 2013, Proxy Statement 2012 Annual Report, www.knighttrans.com.

www.knighttrans.com/Sustainability/index.html.

Murphy, P. R. and D. J. Wood (2011), *Contemporary Logistics*, Chapter 8, 10th ed., Pearson Education/Prentice Hall, Upper Saddle River, N.J.

## Endnote

[1]This section is based on Murphy and Wood, 2011, pp. 246–247.

# 11

# THE IMPORTANCE OF MANAGING INVENTORY

## Introduction

This chapter discusses the potential roles that inventory plays for suppliers, buyers, manufacturers, sellers, and carriers. Specific attention is placed on warehousing's capability to help these supply chain members by managing inventory through the warehouse. Inventory cost factors are outlined, and the implications of good compared to poor management of inventory are explored. This chapter discusses common expectations of the warehouse and the processes employed that influence inventory availability and cost.

## The Importance of Managing Inventory

Partnerships in the supply chain are of great importance for competing in today's business environments. Warehouses are positioned strategically throughout supply chains to facilitate product, information and financial exchanges with materials and other service suppliers, production and manufacturing operations, wholesalers, retailers, and often with the ultimate retail consumer. Interfacing is also critical with carriers upstream and downstream in the supply chain for the management of materials, physical distribution of finished products, and the reverse logistics of products back up the supply chain.

Complex networks of relationships make managing inventory of great importance to all supply chain parties responsible for the making, storing, handling, and financial accounting of materials and products. Corporate objectives align with supply chain logistics management objectives to define the scope of inventory policies and processes for each supply chain constituent. Objectives may vary from one party to the next; however, managers subscribing to the concepts of supply chain management and market orientation can design objectives to achieve quality and value ultimately that transpires into satisfied customers. Market-oriented businesses focus on the wants, needs, and desires of the customer, and in doing so design products and services to meet customers' expectations.

Warehouse operators must manage inventory through their logistics services offered to their customers and manage inventory in ways that help carriers to provide for the customer.

## The Complex Role of Inventory

Now take it from the beginning of the supply chain where Sitka spruce trees are harvested from the Alaskan forest ultimately to be crafted into custom guitars for stage performances or even for the less-expensive beginner guitar packages purchased from discount and online retailers and music stores. Figure 11-1 illustrates the complex stages of a guitar manufacturing supply chain. The authors along with a graduate student conducted research with the manufacturers and their suppliers of four leading guitar brands. Hours of phone call interviews with suppliers, buyers, production managers, brokers, carriers, and retailers helped to create this visual map of a guitar supply chain. Evidence from the figure shows that there is a wide assortment of wood varieties and sources from across the globe called upon to produce the tens of thousands of guitar makes and models.

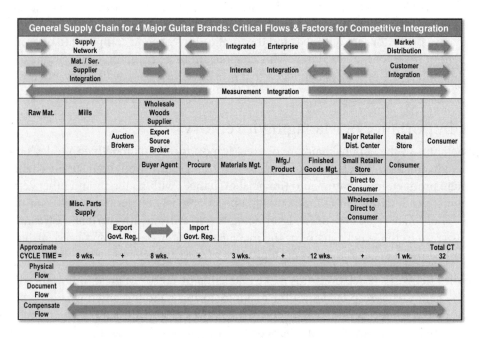

Figure 11-1   Complex guitar manufacturing and distribution supply chain[1]

The lumber company responsible for providing the cutting and transport of the Sitka spruce trees to the lumber mill focuses on gaining economies of scale in loading complete logging tractor-trailers destined for the mill. For the carrier, full loads help to lower the cost of operating the trucking equipment and delivering the spruce logs to the mill. When at the mill, it is logistics responsibility to manage the inventory of logs to efficiently supply the mill that tools the logs and cuts the logs into boards ultimately that will become the pieces of the guitar—among other products that require spruce wood.

The lumber yard can be viewed as a warehouse without a cover for the inbound logs from the field, but after cut and tooled, the output of the mill may consist of several SKUs representing the boards used in the next process for creating the tops of guitars or the billets used for creating the guitar necks. In either case, the completed SKUs are now of greater inventory value per relevant unit, and that translates into greater inventory cost for the warehouse operator and for the mill owner of the product. Moreover, the newly processed and refined product requires a more traditional warehouse roof to protect the wood from precipitation, infestation, and other environmental and critical factors that could alter the quality grade of the lumber.

At this point the mill wants the finished boards and billets to be transported to the next supply chain partner who can further tool the product into finished guitar pieces. However, the objective of the outbound carrier, like the inbound carrier at this level of the supply chain, is to fill up the transportation equipment prior to making the long trip to, for example, the Taylor guitar manufacturing plant located near San Diego, CA. The warehouse must facilitate the storage of the materials and help the carrier to achieve economies of transportation while moving the shipments as quickly as possible so that the mill may transfer ownership of the inventory and the associated inventory carrying costs onto the next party in the supply chain for manufacturing and assembly.

To summarize, the inventory objective of the logging company is to cut and move the logs from forest to mill quickly and with low associated cost. The inventory objective for the carrier is to transport as many logs legally allowable on a single vehicle, so a delay in the transport to obtain a complete trailer load is the preference of the carrier. Production requires a steady flow of logs through the mill process to gain economies of production. Lastly, the mill sales manager and inventory manager want the finished product to move quickly to the next industrial supply chain partner so that an invoice may be cut and the accounts receivable can be collected. The greater number of inventory turns also means a better return on the inventory and warehousing assets utilized to support sales.

The previous description pertains to a single logging supplier, carrier, and mill. Figure 11-1 contains greater detail in depicting how the supply chain becomes significantly more complex when the process requires multiple suppliers providing multiple materials and parts sourced from many locations worldwide.

You can envision how greater the number and variety of supply chain partnerships (nodes or stopping points and links or moving points within the supply chain) comes with equal variety and number of inventory objectives. Assessing the trade-offs between objectives would be necessary to truly manage inventory from a total supply chain perspective.

Warehouse operators must learn to achieve and adjust their own inventory goals while assisting their customers and materials/service suppliers in assessing and meeting their inventory objectives.

As Figure 11-1 illustrates, there is much more to the supply chain of guitar manufacturing and distribution than will be discussed. Finished guitars have a value per unit that exceeds the value of the summated independent raw materials. Warehouses must also help to facilitate the inventory objectives of the wholesale and retail end of the supply chain. At the same time the warehouse must achieve its strategic and operational goals.

Figure 11-2 provides a second example of the critical role that inventory and warehouses play in product life cycle by looking at the supply chain of the Ultra Lightweight Camouflage Net System (ULCANS). The ULCANS is a signature management system used to provide military equipment protection from visual observation, infrared (IR) detection, thermal detection, as well as to provide equipment and personnel shade from the heat of the sun. The manufacturing process for the ULCANS involves numerous suppliers providing raw materials, whereas other suppliers provide value-added services, such as subassembly, to the original equipment manufacturer (OEM). Individual supplier inventory levels are critical to the OEM's capability to manufacturer and deliver the required number of ULCANS to the customer. As can be seen by in the ULCANS supply chain life cycle, insufficient inventory at any one of the supplier locations can cause a perturbation in the process, which, in turn, can negatively affect OEM's capability to deliver systems to the customer.

## General Inventory Categories

Warehouses often have little say in what inventory to stock and when orders for specific SKUs are generated and shipped to customers. A supplier may receive a customer order and immediately pass it to the regional warehouse for filling. The supplier may then assess the remaining stock available for the SKU, combine it with the in-transit stock for the SKU, and evaluate the outstanding orders for the product. This information enables the supplier to determine the amount of stock of product to replenish the SKU to appropriate inventory levels. If the warehouse information stock-keeping records fail to match the physical inventory actually in the warehouse, there is potential for future stockouts on the SKU.

In this instance, the warehouse acts as an inventory custodian for the supplier or customer. The warehouse operator's control over inventory levels, inventory throughput, and inventory turnover is influenced by its capability to fill orders efficiently and ship them according to the customer's required shipping date. Quality of processes, personnel, facility and equipment, and product layout have an impact on achieving this goal. Delayed shipments impact inventory levels and the associated inventory carrying costs.

Inaccuracies in inventory also impact the capability of warehouses to expediently move product from stock to delivery.

## Cycle Stock

Stock ordered routinely to meet anticipated demand is termed *cycle stock*. The economic order quantity (EOQ) was introduced in Chapter 10 to illustrate the interface between carriers and warehouses and how the relationship can influence the levels of cycle stock and safety stock needed. Assuming customers order in relatively predicted cycles, this is the stock necessary to fill orders through the time cycle. The cycle may be determined by a number of factors. Two key factors may include the time necessary to produce the product plus the transit time required to transport the freight to the customer. Suppose it takes 3 days to produce after an order is placed. In addition, it takes 4 days to transport because the carrier utilizes intermodal services to reduce transportation costs. In such a case, the producer would need 7 days of inventory to satisfy customer demand from the time the shipment arrives until the next one arrives.

If demand is known and does not fluctuate, and the carrier arrives on time every time, cycle stock is all that the customer needs to order to meet demand. However, if customers' purchases increase over their expected demand or if a carrier has transit issues, the company would be advised to hold a level of safety stock to meet the wanted customer service levels promised to its customers.

## Safety Stock

Because forecasts, customer wants, and transportation plans may from one time to another fluctuate from the expected, companies must rely on safety stock to uphold the in-stock availability of products. Safety stock and cycle stock work together to help prevent inventory stockouts or shortages. Warehouse space and handling requirements increase as safety stock inflates. Best-case scenario would be for the supplier to evaluate the cost trade-offs between the cost to hold additional levels of safety stock to meet potential customer orders to the costs associated with lost sales if a stockout occurs. Although a one-time lost sale may be easily quantifiable, stockout costs become more difficult to quantify when customers purchase a substitute item, initiate a backorder, or buy the product from a competitor this one time. Of course, the most costly result would be to lose the customer's business forever.

The section describing the iceberg principle goes into greater detail of the warehouse process issues that tend to increase safety stock levels. The combination of cycle stock and safety stock will be utilized to meet predicted demand. Reorder points will be established to trigger an order when on-hand inventory available for orders drops to the point associated with the product of (days lead time required to receive an order) x (daily demand). Should demand increase then safety stock would protect against stocking-out.

## In-Transit Inventory

At first thought, it may be difficult to envision how a warehouse can influence the level of in-transit inventory in the pipeline. Ownership of inventory passes hands as materials and product move through the supply chain. Consequently, at least one member of the supply chain owns the product while it is in motion from suppliers to customers. In-transit inventory, therefore, is as important to manage as in-stock inventory.

Receiving processes at the warehouse determine the ability of the operator to receive the product that is inbound in the most-efficient manner. Delays in receiving, for example, due to poor planning of personnel or equipment needed to service inbound trailers causes stock to remain with carriers for an unanticipated extended time. From a positive perspective, progressive warehouses have instituted procedures for carriers to drop trailers after hours and for some unload and store the physical product in secured dock areas after operating hours.

## Speculative Stock

Not all events influencing customer demand patterns for all types of products can completely and accurately be determined or forecasted. Safety stock helps to alleviate the threat of such unknowns. Events that are anticipated but are not routine and therefore are not covered with cycle stock fall into the category of speculative stock. Marketing and sales, production, and inventory managers may have to speculate the occurrence of many types of events. Some events may pertain to seasonal influences on demand, the anticipation of pricing or interest increases on the horizon, or even the worry of a shortage of transportation capacity availability.

Some examples may illustrate the justification of ordering and stocking inventory above levels common to normal customer demand. Exotic woods often are used in the manufacture of specialized guitar models targeted to customers who purchase for the specific sound resonation and artistic wood grain design that the unique woods produce. Sourcing opportunities are limited for many exotic woods and even for some of the more common woods utilized to make guitars. Woods, for example, sourced from Madagascar have been known to be in short supply when issues native to the region cause breaches in the domestic logistics infrastructure and disrupt the exporting of wood commodities. For this specific reason, one guitar manufacturer purchases and warehouses a 3-year supply of wood primarily grown and sourced from that region of Africa. The company does not want to run out of the wood and be forced to drop the model or line of guitars crafted from it. The warehouse had to identify an appropriate location to store the wood so that the temperature is moderated to keep the correct levels of moisture in the wood during storage.

West coast ports have been shut down due to modern-day longshoremen union strikes that have prohibited the unloading and loading of merchant marine cargo ships. Smart import managers keep up with port activity to protect their firms from potential supply issues. Inventory managers may work with procurement, logistics, manufacturing, and sales to evaluate the forward stocking of speculative stock in anticipation of any major contract negotiations and potential union strikes that may disrupt the flow of imports into the country.

Warehouse and inventory costs increase as buyers procure greater levels of inventory in speculation of future events influencing source availability. Longer term storage of speculative stock may require a warehouse product layout alteration to accommodate the increased amount of inventory on-hand for the SKU. Large quantities held for lengthy periods may be prime candidates for dedicated storage.

## Cost Implications

Materials have a dollar cost affixed to a unit of measure, such as per one hundred pounds (also termed hundred-weight), per cubic or linear foot or yard, per ton, or any other unit size description accepted by the specific industry. Finished goods are also measured in terms of unit value, such as a per case or pallet value, square foot or cubic foot, or again some other well-accepted industry unit measure.

As indicated in the previous chapter, inventory carrying costs may be influenced by several factors. The magnitude of influence may be different depending on the products warehoused. Obsolescence and loss/theft, storage and handling, interest, tax, and insurance are the most commonly quantifiable inventory cost variables.

Although the warehouse may not be responsible for determining levels of inbound stock or outbound order quantities, the operator's policies, procedures and personnel certainly impact the key costs associated with holding inventory.

Product rotation pertains to the warehouse procedures for moving the oldest product out of the warehouse prior to shipping newer product on orders. Although the first-in first-out (FIFO) method may be preferred with products containing sensitive sell by dates or product codes, cross-docking product from inbound to outbound trailers brings opportunity and complexity to proper product rotation. Inventory management systems must help to facilitate cross-docking opportunities when appropriate, but must also protect product from becoming obsolete by determining the most optimal point to merge stored inventory with near code dates with that of cross-docked freight.

Loss and theft of product implies a large assortment of issues. Poorly trained warehouse employees can misplace product even when the most sophisticated inventory tracking and location systems are in place. In one 30,000 square foot warehouse, the manager decided to institute a randomized stocking system. He believed that it would help to reduce receiving time for inbound trailers and railcars. Prior to this the warehouse had dedicated locations for each SKU. After 2 months, the manager noticed that products were being placed back into their traditional locations. He discovered that the employees were overriding the system and changing the slots generated by the optimization algorithm in the warehouse management system (WMS). Although the employees felt that inventory integrity would be best with the fixed locations for each product like in the past, the ability of employees to change product locations within the system provided opportunities for product loss.

Clients are charged for the movement of product and the space occupied by products. Both factors have associated costs with respect to the time required for each. Warehouse operators failing to establish an efficient standardized process for product movement and storage will experience increased labor and equipment costs, and space required for a customer may be greater than anticipated by the warehouse. Increasing costs to manage a client's freight will initiate additional service charges not detailed in the contract and encourage warehouse management and sales to attempt to renegotiate pricing with the customer. Again, the warehouse has the capability to establish processes and training that combat the need to increase rates for customers.

## Forecasts, Sales, and Customer Service Implications

Sales forecasts should include expected base sales that are the sales expected without any additional influences. Cycle stock was previously discussed and includes the amount of product necessary to meet the demand of customers occurring within the cycle time period. This would be the base sales for the SKU or product groupings being forecast. You've seen how warehouse service can influence the availability of stock when ordered. Lost or damaged product or orders that fail to ship on time may reduce the base level of customer demand.

Seasonal fluctuations impact anticipated demand when customers purchase a greater percentage of products above the base sales during specified time periods that tend to recur annually. Although warehouses may not influence this type of demand increase, operations within the warehouse must gear up for increased inbounds to unload, greater variety in SKUs that may or may not have handling history, and greater number of outbounds to meet the specific delivery dates within the season. Assuming the client's base demand is steady and they expect an 18 percent additional seasonal sales, the warehouse can expect 18 percent greater volume to manage. Additional personnel may need to be hired and trained. More equipment may be necessary. Overall warehousing costs can expect to increase, whereas customer service levels are expected to be retained.

Business cycles may also influence demand. Recessions bring upon reductions in customer spending power, and consequently, demand would follow the downward slope. The year 2008 brought on the decline in consumer and corporate spending power. Eventually, logistics companies and operations were stuck with an over capacity of space availability, such as a number of trailers or square footage in warehousing. As demand receded, warehouse operators were forced to scale back on available space because materials and products were not flowing through the supply chain at the same rate and scale enjoyed just prior to the declining business cycle. Inventory levels remaining in the pipeline needed longer term storage, but the valuation of the inventory was also on the decline. Although it is not the point of this discussion to resolve warehousing's role in managing through the decline, it is important for warehouse operators to stay on top of leading indicators of declining or improving business cycles so that they may put into motion appropriate strategic plans and operating strategies.

Marketing and sales promotions influence demand, and expected sales increases from specific activity should be included in forecasts. A manufacturer's production over-run of toilet paper meant a significant cost-savings opportunity for a buyer for one low-cost retailer. The special purchase and accompanying retail promotion was felt by the retailer's distribution center because it had to manage several truckloads of promotional product in addition to its routine throughput. This also meant that carriers seeking inbound appointments for nonpromotional items would have to lay over the weekend to be unloaded on Monday.

Warehouse managers must be familiar with the impact that sales promotions can have on forecasted demand and consequential distribution activity. Collaboration with marketing and sales well in advance of special promotions can serve the retailer, warehouse, and carrier well in maintaining the integrity of the distribution service provided to all and the cost involved.

Although warehouses may receive demand forecast information and data from clients, operators must understand the factors specific to the forecast. It may also be good for the warehouse operator to collect data and perform her own forecast by customer and SKU, and compare forecasts with clients. Forecast methods may include using weighted moving averages; whereby, recent sales over several periods are utilized to predict the next period's demand. The uniqueness is that the most-recent sales data is weighted heavier than the furthest sales period data. This assumes that the closer a demand period is to the forecasted period, the more representative that the period's demand will be to the period being predicted. A more sophisticated weighted approach is performed through a forecasting technique called *exponential smoothing*. The alpha factor is generated and utilized to set weights on the previous forecast based on the resulting difference between the previous forecast and the actual demand realized to predict the next period's demand. Simple and multiple regression analyses and mathematical modeling may also be utilized to forecast demand.

Difficult-to-forecast items require aggregating to a higher level and then creating a more general forecast or relying on the expertise of seasoned managers. Error in the forecast indicate the magnitude of disparity between the forecast and the actual sales. Mean absolute percentage in error takes into account both overforecasting and underforecasting error. An average forecast is created by obtaining the average forecast error without regard to the signage of individual item error.

## The Iceberg Principle

Most everybody knows the story of how ice concealed below the surface of the ocean sunk the Titanic. Although the tip of the iceberg was an indicator of trouble, no one knew just how big the trouble was below the waterline. Warehouse managers can learn lessons from the common industry concept of the iceberg principle. The iceberg represents problems within the warehouse, and between clients, carriers, and receivers, that together cause great inefficiencies and increased costs associated with inventory.

Figure 11-2   The iceberg principle

As can be seen from the previous discussion on forecasting and sales influences on warehouse operations, many positive factors pertain to increased sales that can demand greater levels of inventory. However, warehouse problems, internally or externally, often are covered up when greater levels of safety stock are held so that the problems don't negatively impact customer service levels.

Potential warehouse problems include but are not limited to

- Internal problems

  - Receiving delays cause inventory availability to be less than what actually may be in the warehouse.

  - Mishandled product may equate to lost or damaged product that must be replaced to satisfy customer demand.

- Externally

  - Delayed shipments from the warehouse may cause customers to hold greater levels of safety stock to avoid empty shelves.

  - Product shortages from the warehouse increase safety stock for the receiver.

Although any number of warehousing issues may poorly impact inventory levels, the warehouse manager must monitor inventory levels and accuracy, as well as, monitor discrepancies in orders shipped to identify potential warehouse issues causing the swelling of safety stock in the warehouse or with the customer.

## Errors and Reconciliation

As steward of corporate inventory, such as a private or for-hire warehouse operation, managers must be obsessed with maintaining the integrity of inventory. Errors must be identified and product discrepancies reconciled; otherwise, the level of safety stock will rise to a level unacceptable by the company and clients. Researching inventory inaccuracies will disclose the underlying problems with processes and people that, in turn, must be rectified.

### Cycle Counting

Reconciling inventory discrepancies becomes more difficult the greater the time period expiring from the error to researching and reconciling. Although quarterly, bi-annual and annual physical inventory counts are common for accounting purposes, they do not allow for the immediate or near immediate resolution to problems and reconciliation of inventory. Today's progressive managers look to cycle counting for aiding in identifying inventory problems quickly. Doing so ultimately helps to reduce inventory levels and improve distribution service.

Cycle counting pertains to the physical counting of designated SKUs on an ongoing basis. The basis for determining when to cycle count an SKU may differ between warehouses, but the concept remains to count items that have experienced recent handling activity or those with special characteristics that make them a greater risk for loss or overshipping.

Discrepancies between the physical item count that exists within the warehouse and the count for the SKU on the books can be identified more expediently with frequent counts of highly active products.

Frequency of handling is determined by the frequency of an item ordered or received each day or could be the frequency of an item being moved within the facility during consolidating stock or moving units of an SKU to the OS/D area for rework or repair.

The value of an item may also serve as a basis for including the SKU in a cycle counting rotation. SKUs with greater value would be counted in some cases even if the day's or week's activity did not include that specific item. The value of the item if lost would be of greater concern compared to a lesser valued item.

Cycle counts may also target warehouse sections. Sections containing the greatest activity or valuable SKUs would be counted more frequently than sections without activity or lower-valued item sections. Sections with greatest activity may be counted daily, whereas less-active sections may be counted once per week or monthly.

## Physical Inventory

The comprehensive physical inventory of the entire warehouse provides for the reconciliation of errors that exist between the bookkeeping inventory records and the actual product inside the facility. Similar to a cycle count, it provides warehouse operators and clients a level of confidence in the accuracy of inventory on-hand to meet customer needs and for valuation needs. Physical inventories are more disruptive and require greater resources compared to conducting on-going cycle counts. Conducting a comprehensive physical inventory also requires specific planning ahead of time to ensure that no disruptions to service and deliveries occur during the conduct of the physical count.

With the ever-increasing application of radio frequency (RF) technology, leveraging the accuracy power of traditional bar codes and the more advanced radio frequency identification (RFID) tags, inventory control managers within many warehouses are better equipped to maintain the integrity and accuracy of product in inventory. Technology has helped to prevent warehouse errors during putaway, replenishment, picking, and loading. Reduced errors means that cycle counts and physical inventories can be more efficient and more accurate because there will be fewer inconsistencies between inventory records and actual stock prior to initiating physical counts.

Physically counting and auditing the total stock occupying the warehouse requires procedural planning, identifying and training proper personnel for the jobs of counting, maintaining the records, and reconciling. For some warehouses physical counts may be accomplished with RF readers; however, others may require the use of a tagging system. In either case, managers are encouraged to conduct independent counts for each unit position in the facility. Two independent counts conducted by different counting teams must then be compared to identify any product locations where the two counts differ. A third count is required when the first two counts don't agree. Reconciliation of the final counts must be made with the inventory appearing on the computer records.

It is important to completely segregate inbound product received after the start of a physical inventory from the area designated for counting. Equally important is to keep the inbound from appearing on the books that will be used to compare with the physical stock. Because the OS/D or rework area potentially requires more work to count product because the area may contain multiple SKUs in varying states of repair, it should be scheduled first.

Upon completion of the physical inventory and reconciliation of discrepancies, the inventory on record must be adjusted to match the physical count. Clients and warehouse operators are encouraged to be explicit when writing the physical inventory policy within the contract agreement. Tolerance levels for inaccuracy should be stated, and penalties must be clear for when inventory adjustments to the records exceed tolerances.

## Summary of Key Points

Inventory holds a special place in the supply chain of a product. Without the appropriate amount of inventory, end-product deliveries to the customer are not possible. Yet, excess inventory equates to higher costs and inefficiencies. Management of inventory throughout the supply chain is a salient factor that determines success. Synchronizing the type of inventory to maintain to the product and environmental factors surrounding the product is imperative. A strong link between the warehouse manager and the customer enables appropriate forecasting to be conducted and prevents stockouts from occurring. Conducting physical inventories at the appropriate frequencies help to ensure accuracy and also assist in preventing stockouts.

## Key Terms

- Accounts Receivable (A/R)
- Cross-Docking
- Cycle Counting
- Cycle Stock
- Demand Forecast
- Distribution Center (DC)
- Downstream

- Economies of Production
- Economies of Scale
- Economies of Transportation
- First-In First-Out (FIFO)
- Forward Stocking
- Hundred-Weight
- Iceberg Principle
- In-Stock Inventory
- In-Transit Stock
- Inbound Carrier
- Intermodal
- Inventory Carrying Costs
- Inventory Integrity
- SKU
- Inventory Management System
- Inventory Objective
- Upstream
- Inventory Stockout/Stockouts
- Inventory Throughput
- Inventory Turnover/Turns Invoice
- Obsolescence
- Outbound Carrier
- Physical Inventory
- Product Codes
- Product Rotation
- Reverse Logistics
- Safety Stock
- Seasonal Fluctuations

- Speculative Stock

- Stockout Costs

- Supply Chain

- Trailer Load (TL)

- Transit Time

- Value-Added Services (VAS)

- Warehouse Management System (WMS)

## Suggested Readings

Donald J. Bowersox, David J. Closs, and Theodore P. Stank, Michigan State University, (1999), "21st Century Logistics: Making Supply Chain Integration a Reality," Council of Logistics Management, Oak Brook, IL., p. 29.

Coyle, J. J., C. J. Langley, R. A. Novack, and B. J. Gibson. (2013), *Supply Chain Management: A Logistics Perspective*, 9th ed., Chapter 7 and 9, South-Western/Cengage Learning, Mason, OH.

Narasimhan, R. and S. W. Kim, (2001)"Information System Utilization Strategy for Supply Chain Integration," *Journal of Business Logistics*, Vol. 22, No. 2: pp. 51–75.

Scott Keller and Robert Saxer (2013) "Complexity in Managing Guitar Manufacturing Supply Chains," working paper.

Teresa M. McCarthy and Susan L. Golicic, (2002) "Implementing collaborative forecasting to improve supply chain performance," *International Journal of Physical Distribution & Logistics Management*, Vol. 32 Issue: 6, pp.431–454.

## Endnote

1    Donald J. Bowersox, David J. Closs, and Theodore P. Stank, Michigan State University, "21st Century Logistics: Making Supply Chain Integration a Reality," 1999, *Council of Logistics Management*, Oak Brook, IL., p. 29; Scott Keller and Robert Saxer (2013) "Complexity in Managing Guitar Manufacturing Supply Chains."

# 12

# SELECTING WAREHOUSE LOCATIONS

## Introduction

The location or placement of a warehouse is both a strategic and operational decision that directly impacts customer servicing aspects and cost and price factors. You must consider and analyze many parameters to optimize a warehouse location. This chapter enumerates these factors and their impact on quality, timeliness, and cost. In addition, this chapter provides two warehouse location analysis examples.

## Selecting Warehouse Locations

From the earliest of times, cities have been located near waterways and major crossings where people can sustain themselves and supplies can be reached. A means of transportation facilitates successful migrations and the economic development of communities. Along with such progress, storage and handling facilities are required to maintain supplies for sustenance or for merchant sales.

It is no surprise that transportation plays such a key role in today's decisions for locating warehouses, distribution centers, and cross-docking operations. Chicago stockyards were successful because transportation could now transport livestock from the countryside via railroad to the yards in Chicago. Processors could then prepare meat for refrigerated (ice in the earliest years) transport to further away retail sales and consumption locations.

Stocking points from fields to consumption facilitate marketing exchanges with near and distant markets. Warehouses play a critical role in reaching markets. Transportation availability, quality, and pricing influences the overall total operating cost for the client and operator, and should be considered when locating warehouse facilities.

Other critical factors include, for example, area zoning regulations pertaining to variances and the specific distribution centers' (DC) activity performed within and outside the building, taxation policies, availability and cost of land, labor and utilities, and of course, the anticipated flow of inbound and outbound shipments and the transportation spend associated with these activities.

## Primary Factors

One warehouse located in Olive Branch, MS, received substantial tax incentives and low interest rates to locate and employ area workers. Although not the only consideration, similarly, a completely flow-through distribution center received financial incentives from the state and local governments in central Alabama to construct the building and train employees.

Inventory may be taxed on the dollar value of the stock on-hand at a designated evaluation time. State and local tax laws governing the level of inventory taxation must factor into a company's location decision. Total distribution cost may be lower by locating a field warehouse 100 miles further away from the market if the area has little inventory tax requirements compared to an alternative warehouse within the city limits of the market but having significantly elevated inventory taxes. The familiar trade-off analysis of transportation and inventory carrying cost must include anticipated tax dollars when computing the total cost comparisons between location decisions.

Zoning laws must also be well understood so that the work performed conforms to the restrictions placed on the warehouse property. Although state, local, and county regulations may allow for the storage of a specific product, there may be ordinances restricting the size and weight of trailers entering the zone. Time restrictions may also dictate operating hours available to perform the duties required of the warehouse clients or tenants. Covenants of the location may restrict the nature of distribution and warehousing business that can be conducted and the characteristics of the size, height, and overall footprint of the buildings and trailer lot/staging areas.

For the Alabama flow-through DC, the company received the land for free as incentive for locating in the rural area. Territory for the distribution center included Birmingham, where land availability was limited to accomplish the company's supply chain financial and distribution goals. Location of the distribution center was near a major north-south interstate highway, and the number of employees needed for the operation was easily obtained from the surrounding rural population labor pool.

One other critical criteria is the location of the warehouse or distribution center to the materials or supplier source in comparison to the location of the customer. Complexity is magnified as the DC supports multiple suppliers and multiple customers and markets.

## Facility Location Analysis

Sophisticated location analysis models are designed with the capacity to conduct the complex computations required when many and complex variables enter into the decision. Potential variables include inbound freight volumes and transportation costs. Inbound freight costs can vary by client and SKU if, for example, the warehouse stores raw materials for production, receives finished goods for distribution, or both.

Rail sidings may be required to receive rail carloads of materials for input into a manufacturing production process. Manufacturers must consider the impact of the production process on the weight and consistency of the finished product. Does the product gain weight during production that could cost more in freight transportation bills? If so, the plant operator may be advised to consider a geographic location near the customer or market. However, if product weight loss occurs during production, the finished product would be lighter causing freight rates, based on weight or tonnage, to draw the plant location toward the raw materials suppliers. Private warehouses supporting manufacturing may locate near the plant, and consequently, be influenced by product weight gain or loss.

Table 12-1 provides an example of a weighted average warehouse location analysis. Because transportation costs influence a large percentage of total logistics cost, it makes sense to calculate the best location for a warehouse or DC that minimizes transportation dollars spent on managing inbound and outbound shipments.

Table 12.1: Weighted Average (Rate and Ton-Miles) Warehouse Location Analysis

| Supplier | Rate $ per Ton-mile | Tons | Grid Coordinate Horizontal | Grid Coordinate Vertical | Calculation Horizontal | Calculation Vertical |
|---|---|---|---|---|---|---|
| S1 | 2.50 | 500 | 350 | 200 | 437,500 | 250,000 |
| S2 | 2.15 | 500 | 400 | 500 | 430,000 | 537,500 |
| S3 | 1.80 | 800 | 275 | 500 | 396,000 | 720,000 |
| S4 | 2.50 | 700 | 200 | 420 | 350,000 | 735,000 |
| | | | | Sum | 1,613,500 | 2,242,500 |
| **Markets** | | | | | | |
| M1 | 2.00 | 380 | 600 | 300 | 456,000 | 228,000 |
| M2 | 1.90 | 320 | 600 | 600 | 364,800 | 364,800 |
| M3 | 2.30 | 400 | 550 | 700 | 506,000 | 644,000 |
| M4 | 2.50 | 250 | 700 | 650 | 437,500 | 406,250 |
| | | | | Sum | 1,764,300 | 1,643,050 |
| | | | | S/M grid | 1,613,500 | 2,242,500 |
| | | | | Summation | + 1,764,300 | + 1,643,050 |
| | | | | Grand Sum | 3,377,800 | 3,885,550 |
| *See calculation below for 8,428 | | | | | 3,377,800 ÷ 8,428 | 3,885,550 ÷ 8,428 |

**Supplier:** (2.50 x 500) + (2.15 x 500) + (1.80 x 800) + (2.50 x 700) = 1,250 + 1,075 + 1,440 + 1,750 = 5,515

**Market:** (2.00 x 380) + (1.90 x 320) + (2.30 x 400) + (2.50 x 250) = 760 + 608 + 920 + 625 = 2,913

**\*Summation:** 5,515 + 2,913 = 8,428

Grid for Location of Facility (Coordinates Calculated in Previous Analysis)

| | 100 | 200 | 300 | 400 | 500 | 600 | 700 | 800 |
|-----|-----|-----|-----|-----------|-----|-----|-----|-----|
| 800 | | | | | | | | |
| 700 | | | | | M3 | | | |
| 600 | | | | | | M2 | M4 | |
| 500 | | S3 | | S2 | | | | |
| 400 | | S4 | S1 | *(401/461) | | | | |
| 300 | | | | | | M1 | | |
| 200 | | | | | | | | |
| 100 | | | | | | | | |
| 0 | 100 | 200 | 300 | 400 | 500 | 600 | 700 | 800 |

*Indicates grid coordinates and approximate location of facility location.

Step 1 would be to identify the location of suppliers that will be channeling freight into the DC. The locations may be mapped on a grid with vertical and horizontal axes representing miles from a zero distance starting point where the X and Y-axes intersect. For step 2, markets or customer locations would then be placed on the grid according to their individual distances from the zero starting point. Step 3 would entail estimating the total tonnage expected to move from supplier to the warehouse. Do this for each supplier lane. Tonnage estimates would also be needed between the warehouse and each market. This would be step 4 in the process.

Transportation rates per ton-mile would then be calculated by multiplying the transportation rate per ton-mile for supplier S1 (2.50) to the tonnage for the product being shipped from the supplier (500). To factor in the miles, it is necessary to multiply the result 1,250 by the distance of the supplier from the zero intersection-point for both the X-axis and Y-axis. The result utilizing supplier 1's (S1) horizontal grid coordinate is 437,500 (1,250 x 350 = 437,500). Do the same for S1 utilizing the supplier's vertical grid coordinate (1,250 x 200 = 250,000). The same procedure must be done for each supplier and each market, and then summate the supplier and market horizontal column totals (1,613,500 + 1,764,300 = 3,377,800) and then the vertical column totals (2,242,500 + 1,643,050 = 3,885,550). The final two steps involve calculating the product summations for rate multiplied by tons for all suppliers and markets (See the following calculations).

**Supplier total**:

(2.50 x 500) + (2.15 x 500) + ($1.80 x 800) + ($2.50 x 700) = 1,250 + 1,075 + 1,440 + 1,750 = 5,515

**Market**:

($2.00 x 380) + ($1.90 x 320) + ($2.30 x 400) + ($2.50 x 250) = 760 + 608 + 920 + 625 = 2,913

***Summation**: 5,515 + 2,913 = 8,428

Table 12-1 indicates the finalized grid coordinates for locating the warehouse to minimize the total inbound and outbound transportation costs. The last two rows in the table illustrate the final calculations for the horizontal grid location (3,377,800 / 8,428) and for the vertical grid location (3,885,550 / 8,428). Based on the estimated total ton-mile costs for servicing suppliers and markets, the warehouse or distribution center would best be located at the intersection of horizontal coordinate 401 and vertical coordinate 461. Minor changes in volume or rates do not significantly impact the geographic location resulting from utilizing the weighted average (rate and ton-miles) warehouse location analysis.

Transportation costs may be the dominating factor in locating some warehouses. However, for others, cost trade-offs must be assessed between transportation costs, warehousing costs, and inventory carrying costs. Total cost comparisons must be made to assess the viability of one warehouse location option to another.

## Co-Locating a Home Healthcare Agency and Home Medical Equipment Company: An Example from the Front Lines (Section Based on an Interview with Maureen McBride, Home Healthcare Professional)

A Florida-based home healthcare agency, called FHH, was looking to expand its market reach by obtaining operating authority in a six-county area new to the company. The agency provides quality care for patients who are discharged from hospitals so that patients can continue healing in the comforts of their home. Patient healthcare costs are substantially reduced when care can be provided at home compared to the cost if the patient remains in the hospital facility or is admitted to a skilled nursing facility.

Continuing with the example, FHH receives patient referrals from hospitals but also from physicians' offices, nursing homes, assisted-living facilities, and skilled nursing facilities. When FHH receives a patient referral, an assessment of the patient is conducted by a medical clinician. This is conducted in the home of the patient, and based on the physical assessment of the patient, the appropriate clinician (nurse, physical therapists, occupational therapist, and so on) schedule visits to assist the patient in the healing process. FHH deploys agency salespeople to all the locations previously mentioned. Professional sales agents establish and retain relationships with referral sources and monitor the care provided to the patient.

Location selection for the new FHH office required the mapping of the referral sources and an analysis of the population clusters for adults 65 years and older. FHH specializes in providing medical care to this segment of the healthcare market.

Corporate officers decided that it would be a good sales and operations strategy to co-locate FHH with its sister company, Florida Home Medical Equipment (FHME). Locating them under the same roof would enable the integration of the businesses to gain sharing opportunities and efficiencies. Building costs, personnel, and office technology and equipment would be shared. Moreover, the customer service goals of both FHH and FHME would be enhanced because FHH field sales personnel and clinicians could assist in delivering equipment to patients if they are scheduled to visit a patient who needs medical devices from walkers and canes to oxygen tanks. Along with gaining cost and service economies for the building and personnel, both companies can help to promote each other's services to referral sources. Hospitals and doctor's offices often prefer to deal with a single source for their patients' home care and equipment needs. One call can allow the referral source to schedule services and products from FHH and FHME.

As can be seen from the example, locating an office central to the medical facilities serviced can enable the FHH sales staff to reduce the cost of travel (including vehicle expenses and travel time) and enable an increase in the quality time promoting their services and visiting patients: two key components of successful selling in the home healthcare and equipment business. FHME will benefit equally from operating its private warehouse and equipment distribution services from a location central to the same referral sources. A mileage-grid analysis similar to Exhibit 12.1 would help the companies to identify the central location/area.

Both FHH sales personnel and FHME equipment delivery personnel would further benefit from an analysis to evaluate the potential routes from the proposed central location to the medical facilities serviced. Although both have vehicle operating costs and compensation costs that need to be considered when weighing alternative facility locations and traveling routes, the equipment company incurs the added expense of in-transit inventory cost.

Hospital-type beds delivered to the home can cost more than $500, whereas a single cylinder of oxygen can amount to $25 or greater. Oxygen on demand (O2 concentrator) is gaining traction in the industry, whereby, a device takes the oxygen in the air and filters it creating pure oxygen in the patient's home. Unit prices for a standard concentrator start at approximately $500 and increases to a price of greater than $1,800 for a portable O2 concentrator, but, the variable cost is substantially lower than the O2 cylinder price. FHME owns the inventory in stock and in transit. Durable medical equipment has a high per unit inventory carrying cost, and therefore, inventory reorder points are set as low as possible, and replenishment transportation services must be highly responsive. A one-time stockout of an oxygen cylinder for a home-bound patient could be life-threatening, and stocking out when a hospital orders O2 can jeopardize the relationship between FHME and the medical referral source. For this reason, the safety stock level for oxygen cylinders would be set higher than for *bent metal*, an industry term used to describe less-expensive aluminum walkers, canes, and bedside toilets.

Physicians may refuse to write patient orders for an O2 cylinder delivery from an equipment company that is greater than, for example, a 1-hour drive from the hospital. The physician does not want to risk a delayed delivery due to unforeseen delivery issues. This adds complexity to facility location when drive time, including heavy traffic delays, must be factored into the location analysis. For this reason, alternative routes should be evaluated for the efficient transit time of salespeople, clinicians, and equipment delivery personnel.

Routing models exist to evaluate equipment delivery. Routing costs should be assessed from the point of each potential or proposed facility location. Table 12-2 provides an example of one alternative routing solution when factoring in distances between the co-located FHH/FHME medical staff office and equipment warehouse and the medical facility referral sources. The distance between the origin and all the service locations must be established. In the example, FHH/FHME is located 6 miles from the nearest medical facility client; a skilled nursing facility (SNF-1). Referral source ALF-3 (assisted living facility #3) is located the furthest distance from the origin agency and warehouse location. Capacity requirements for each of the customers is listed in column two. For example, the hospital labeled HOSP-1 requires on average 30 square feet of truck space for the medical equipment to be delivered. In this example, the maximum amount of available space on the truck for medical equipment is 115 square feet. The numbers are simplified for illustration purposes.

Table 12-2: Home Healthcare Sales and Medical Equipment Routing Example

| Referral Source | Capacity Required (Sq. Ft. Demand) | MILES | | | | | | | |
|---|---|---|---|---|---|---|---|---|---|
| | | FHH/FHME AGENCY | SNF-1 | ALF-1 | HOSP-1 | ALF-2 | HOSP=2 | ALF-3 | HOSP-3 |
| FHH/FHME AGENCY | Origin | 0 | | | | | | | |
| SNF-1 | 16 | 6 | 0 | | | | | | |
| ALF-1 | 20 | 8 | 7 | 0 | | | | | |

| Referral Source | Capacity Required (Sq. Ft. Demand) | MILES | | | | | | | | |
|---|---|---|---|---|---|---|---|---|---|---|
| HOSP-1 | 30 | 9 | 10 | 10 | 0 | | | | | |
| ALF-2 | 26 | 7 | 18 | 15 | 7 | 0 | | | | |
| HOSP-2 | 30 | 18 | 25 | 22 | 8 | 9 | 0 | | | |
| ALF-3 | 20 | 25 | 30 | 27 | 15 | 12 | 8 | 0 | | |
| HOSP-3 | 40 | 18 | 24 | 26 | 14 | 9 | 8 | 7 | 0 | |
| Max. Truck (115 Sq. Ft.) | | | | | | | | | | |

When considering the miles from the origin agency or warehouse to the nearest customer, and continuing on to the next nearest customer, a route alternative could be established for a salesperson. When constrained by the square footage needs for each delivery, a two-truck delivery solution results (see Figure 12-1). This is exhibited in the routing diagram below the table. Equipment delivery truck #1 will be assigned the following route from the FHH/FHME (originating equipment warehouse):

**Stop 1**: Skilled nursing facility (SNF-1)

**Stop 2**: Assisted-living facility (ALF-1)

**Stop 3**: Hospital 1 (HOSP-1)

**Stop 4**: Assisted-living facility (ALF-2)

**Stop 5**: Return to FHH/FHME (originating equipment warehouse)

Delivery truck #2 would be needed to supply the additional customers because of the square foot capacity constraint (115 square feet) available for transporting material on each truck. Ninety-two square feet of space will be required for the equipment transported in the first delivery truck to customer stops 1 through 4 (16 + 20 + 30 + 26 = 92) or approximately 80 percent capacity utilization (92/115). Adding the next-in-line customer stop would exceed the capacity availability in the first truck. The remaining customer locations would be serviced by the second delivery truck: Stop 1: HOSP-2 (30 square foot needs); Stop 2: ALF-3 (20 square foot needs); and Stop 3: HOSP-3 (40 square foot needs) totaling 90 square feet or approximately 78% capacity utilization (90/115).

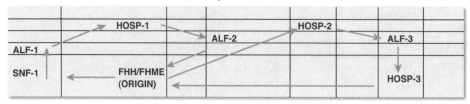

Figure 12-1    Two-truck delivery route alternative

It is clear that there are other important factors that must be evaluated by the FHH and FHME before finalizing a location decision. For a city of approximately 200,000 people, the zoning manual exceeds 300 pages of details. State, city, and county regulations must be well researched prior to selecting a warehouse location. Following are some of the additional details that had to be worked out by FHH and FHME prior to final location approval:

- Location zoning had to be approved for retail sales because some of the medical equipment would be sold to the general retail public.

- A city business license had to be retained.

- The state required a letter from city planning and zoning approving the site for the medical business that would be conducted at the facility.

- A state license from the Agency for Healthcare Administration had to be retained.

- A state bio hazardous waste license had to be retained because the facility would be handling medical waste. Identification and the amount of the hazardous waste was required.

- Oxygen storage and a safety license was required to store oxygen cylinders.

- Level II background screening includes employee Social Security number, height, weight, eye color, and fingerprint to identify any violent arrests/background. At a cost of $50 per employee, the screening is required because the medical and equipment staff interact with "population at risk" or patients 13 and younger and those 65 and older that may be easily taken advantage.

Whether medical equipment supplies or stuffed toy animals, warehouse site location requires a complete understanding of the market and population characteristics, product and service characteristics, geographic factors, and safety and economic regulations. Locations nearer similar existing corporate facilities may make it easier to share resources especially during initial start-up of the new facility. Experienced employees can be utilized to service the new referral sources, as well as, train new employees for long-term employment within the new facility. In the case of delivering sensitive medical supplies such as oxygen canisters, alternative routings are critical. For example, Panama City Beach, Florida, has three main roads running east and west: front beach, middle beach, and back beach roads. During the height of the tourist season, easy access to the alternative routes is essential to avoid long delays in transit time to deliver critical medical equipment.

## Summary of Key Points

Warehouses play a critical role in reaching markets. Availability of transportation networks plays a key role in today's decisions for locating warehouses and distribution centers. When determining the location for warehouses, cost trade-offs must be conducted and analyzed and include transportation costs, warehousing costs, and inventory carrying costs. Total cost comparisons must be made to assess the viability of one warehouse location option to another. Warehouse site location requires a complete understanding of the market and population characteristics, product and service characteristics, geographic factors, and safety and economic regulations. Depending on the product, warehouse location decisions may not impact only customer service and operating costs, but they may also have life and death implications.

## Key Terms

- Area Zoning Regulations
- Distribution Centers (DC)
- Financial Incentives
- In-Stock Inventory
- In-Transit Inventory
- Routing Costs
- Inbound Transportation Cost
- Inventory Carrying Costs
- Safety Stock
- Inventory Reorder Point
- Inventory Taxation
- Private Warehouse
- Outbound Transportation Cost
- Stockout

- Supply Chain

- Tax Incentives

- Total Distribution Cost

- Transportation Costs

- Variable Costs

- Variances

## Suggested Readings

Akerman, K. B. (1997, 2012), *Practical Handbook of Warehousing*, 4th ed., Chapter 11, Chapman and Hall, New York, NY.

Korpela, J., A. Lehmusvaara, and J. Nisonen, (2007), "Warehouse operator selection by combining AHP and DEA methodologies," *International Journal of Production Economics*, Vol. 108, No. 1–2: pp. 135–142.

Min, H. (1994), "Location Analysis of International Consolidation Terminals Using the Analytical Hierarchy Process," *Journal of Business Logistics*, Vol. 15, No. 2: pp. 25–44.

ReVelle, C. S. and H. A. Eiselt, (2005), "Location analysis: A synthesis and survey," *European Journal of Operations Research*, Vol. 165, No. 1: pp. 1–19.

Vlachopoulou, M., G. Silleos, and V. Manthou, (2001), "Geographic information systems in warehouse site selection decisions," *International Journal of Production Economics*, Vol. 71, No. 1–3: pp. 205–212.

# 13

# SAFETY AND SECURITY

## Introduction

There is no higher priority within warehouse operations than the safety of personnel, followed closely by the security of customer goods. The inherent and dynamic nature of the myriad of activities, often occurring simultaneously, within the distribution center creates an ever-present and always potentially dangerous operating environment. Creating a safe environment begins with the person in charge of the warehouse operations. That person must constantly work to create and maintain a culture of safety. Processes must be put in place, and just as important, personnel must be trained to use and improve safety processes and procedures to ensure a safe work environment. From designated product placement to safe product movement, all aspects of daily operations must be designed with safety in mind. Numerous safety cues are endemic to the warehouse, and all employees must be taught to use all their senses to recognize these cues. Securing the product from environmental threats and from handling damage or theft is also of paramount importance. Safety and security within warehouse operations are two fundamental aspects that must be accounted for from the beginning.

## Safety and Security

While working as a frontline merchant marine terminal supervisor in Long Beach, CA, one of the authors recalls overhearing a conversation between the super cargo (lead union clerk) and a utility tractor (UTR) driver. The UTR driver had just won a substantial amount of money by playing the Lottery. Although the winnings were enough to support a comfortable retirement for the UTR driver, he continued to show up for work at the 600-acre shipping terminal. The super cargo encouraged the UTR driver to retire from the waterfront saying, "It's too dangerous to work down here if you don't have to."

The terminal managed the discharge and loading of some of the world's largest container, general cargo, bulk, and steel ships. Working multiple ships sometimes required more than 100 longshoremen including UTR drivers, clerks, crane and heavy lift operators, mechanics, general purpose workers, and supervisors for the ship and dock. The operation was more complex when terminal operations included interchanging the containers drayed to and from the port by motor carrier drayage companies. Import steel slabs were also offloaded from the ship and loaded onto flat cars that were staged for a shortline rail land bridge to a steel mill within the region.

The terminal consisted of three large general cargo transit sheds for temporary storage and cross-dock freight, two massive container yards, a bone yard or empty chassis yard, and a six-to-eight lane interchange receiving and shipping gate. Many times the yard would be rearranged to best accommodate operational efficiencies for the marine work and dock work. General cargo would be staged in varying configurations within transit sheds and throughout the open dock space. Temporary storage for overflow equipment and cargo rotated wherever the space would allow. The terminal and cargo layout was ever-changing. During the time the co-author was working for the marine terminal, there were three accidents resulting in deaths for even the most experienced workers. Unexpected shifting of staged cargo, movement of freight while loading, and driving through an area in which temporary equipment was stored were the conditions relating to the three accidents. Conditions such as these often exist when storing and handling freight in warehouse and distribution operations whether within an enclosed or open storage area or within the lots where vehicle entry, exit, and parking takes place. Figure 13-1 illustrates the discharging of ocean freight containers from vessel to wheeled chassis. The ship pictured can contain containers above deck and break-bulk freight below deck. Discharged containers can be drayed to a cross-docking facility where the contents are unloaded and placed on alternate trailers for delivery to customers. General break-bulk cargo are staged outside or inside transit sheds awaiting pickup from delivering carriers.

Figure 13-1   Discharging containers from ship to chassis

Along with the safety of personnel, warehouse operators must provide for the protection and security of product flowing through their facilities. When storing oxygen cylinders for medical patients, operators must take care to place the cylinders away from heat. Regulations require these cylinders to be located away from high-wattage lighting. Some facilities are equipped with containment storage rooms containing walls and doors thick enough to withstand the pressure of a compressed liquefied or gaseous material tank/cylinder exploding and projecting into the structure. The containment room protects the product by keeping it away from adverse conditions, and protects personnel and other product outside of the room from potential cylinder projection.

Safety is one of the primary concerns of managers supervising warehouse and distribution operations. Key measurements of safety must be designed and results compared with safety goals so that continuous improvement of safety for personnel and product can be achieved.

## Preventing and Reducing Warehouse Accidents

Most important, accidents can be avoided with proper storage processes, employee training, supervision, and warning indicators in place. Whether an open or enclosed product storage area, safety of personnel and inventory must begin at the building or selection of a warehouse facility. Characteristics of the building, materials handling and storage devises, product flows inbound and outbound, inventory flow within the facility, and even areas where product dwells within or outside of the facility are important to consider when selecting a site for storage and distribution.

Specific to warehouse and storage (North American Industry Classification System or NAICS, sector 4931), the Bureau of Labor Statistics (BLS.gov) indicates that in June 2013, there were 688,800 (seasonally adjusted) workers employed in this business sector. Figure 13-2 indicates that the number of fatalities among workers in warehousing and storage has declined over the past 3 years. In 2010, the number of workplace fatalities reached 20 or a 17percent increase over 2009 reports. However, in 2011 (16 incidents resulting in fatalities) and 2012 (12 incidents resulting in fatalities) the rate has been dropping for the industry sector.

Similarly, the nonfatal rate of injury and illness for employees in warehousing and storage has declined from 5.9 cases in both 2009 and 2010, to 5.5 cases per 100 full-time employees (FTE) in 2011. Cases involving days off work, job restrictions, or transfer of duties declined, as well (2009, 4.3 per 100 FTE; 2010, 4.1 per 100 FTE; and 2011, 3.8 per 100 FTE).

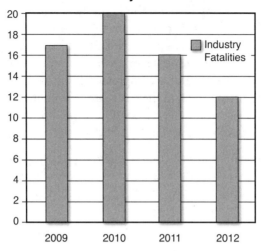

**Industry Fatalities**

Figure 13-2    Number of fatalities among warehousing and storage workers
(Bureau of Labor Statistics, BLS.gov)

Although the trends for the industry segment are positive, managers and frontline ware-house employees must exercise vigilance when working within industrial-type environ-ments requiring heavy equipment to move and store volumes of heavy freight.

## Visual Safety Communications' Strategies and Systems

A good place to begin this discussion is with the receiving of freight at the warehouse dock door. Buildings equipped with red/green light trees protect warehouse employees by indicating safe or unsafe access conditions specific to a receiving or shipping door-way. Similar to a traffic light, with one red (stop) light and one green (go) light, each warehouse doorway would have a red/green light tree secured to the inside wall near the doorway and another red/green light tree secured to the outside wall near the doorway. A driver knows that the doorway is safe to back into or out of when the outside light is green. Red outside lights warn drivers to stop; don't enter or exit the doorway with your trailer. The inside red/green light tree is utilized in the same way so that when a red light appears by the dock door, a warehouse employee knows that it is unsafe to enter a trailer parked in the doorway. After the carrier safely secures the trailer to the dock and door-way, and the trailer is ready for loading or unloading, an inside green light signals to the lift driver that it is safe to enter into the trailer. At this point the red light appears outside the doorway indicating to drivers that it is not safe to move the trailer.

## Protecting Workers and Equipment: Dock Levelers, Restraints, and Bumpers

Dock bumpers should be installed at every warehouse freight dock door to prevent trailers from direct contact with warehouse walls that are not designed to absorb the impact of a 90,000-pound load and tractor/trailer. The integrity of the doorway is maintained, while carrier equipment and driver are safer.

When the trailer is positioned within the dock doorway, dock levelers are designed to bridge the gap between the doorway and the trailer opening and thus creates a continuous surface for a lift, worker, or conveyor to enter and exit the trailer. Although there are many configurations of products sold for this purpose, perhaps one of the safest is a hydraulic dock leveler that removes the need for a forklift to position a dock board or plate for use and remove or store it when not in use. Specialized ramps are also utilized to level or bridge the opening between the surface of the loading dock and that of railcars. For motor carriers, securing and stabilizing a trailer to the dock may be accomplished with a latch or steel hook devise designed into the dock leveler system that hooks to the trailer's steel structure, thus, holding the trailer in place and resisting its movement. Drivers should always utilize wheel blocks as a redundant safety measure.

## Safe Product Movement

Some lifts enable the unloading of double-stacked pallets, and single-double lifts enable the simultaneous handling of two side-by-side pallets. Both would be utilized to gain efficiencies in product movement from trailers to storage and in loading. By nature of the handling process, moving forklifts and palletized product create an environment in which employees must remain aware and exercise caution. Pedestrian walkways inside the warehouse should be designated with highly visible floor paint, as well as signage to indicate safe walking paths within the warehouse. However, forklifts may not always remain within a routine traffic flow while moving freight. Actually, they may take irregular routes throughout the facility depending on the location of the product and trailer activity. Adding to the complexity, warehouse activity may fluctuate depending on the number of scheduled receipts and shipments, and any other necessary housekeeping duties, such as repositioning overages and damages or handling freight that fits into the distressed category (that is, requiring inspection for determining disposition).

Licensed forklift operators must constantly lookout for people that may be unaware of his presence and movement. Inbound and outbound freight volumes fluctuate daily. At one moment freight may occupy a space and the next an employee or piece of equipment may occupy that same location.

However, an empty space on the dock may quickly be filled with product causing a net loss in visibility in that section of the dock. Equipment variety shouldn't present danger; however, when operating manually assisted pallet jacks within the same area of heavier and faster moving forklifts, operators and clerks must be aware of the presence of all equipment sizes. One minute an AGV may occupy the space where previously a lift or even a facility bicycle for transporting a worker recently occupied. Some equipment compared to others within the same workspace simply have better visibility and provide clearer visibility when operating than others. Additional factors that may impact safety when checking in freight and moving it to and from inventory storage include

- Receiving clerks may be inspecting freight for count and condition.
- Close quarters might exist between pallets of freight and the aisles where forklifts move.
- Multiple lifts might operate within the same dock space.
- Movement of palletized freight may have damaged cartons causing a breach in the integrity of the load and cause it to shift and topple.
- Time pressure exists to receive and putaway freight.
- Operating forklifts around corners and within narrow aisles might occur.
- Lifting palletized freight into high-bay storage slots might occur.

While visiting a newly constructed and equipped confections distribution center, one of the authors asked the DC manager why the aisles were so wide; when the trend seemed to be narrow aisles in favor of greater product storage capacity. The manager explained that the DC wanted to improve travel time while putting away and retrieving product, and the wide aisles allow for the forklifts to operate at the highest safe-speed designed into the equipment. Generously sized end-of-aisle oval mirrors allowed for quick and easy visual warning of any on-coming lift or personnel traffic nearing the aisleway inter-section. The strategy worked. Increased safe forklift speeds enabled the improvement of putaway and picking processes to the extent that inventory velocity and inventory turns improved. Inventory availability improved because receipts were put away and recorded quicker. Order cycle time was improved because the fulfillment process required less time to complete and load a shipment.

The net result included needing reduced inventory levels to service existing customer demand. This equated to more efficient utilization of available space. Managers must carefully consider the goals of warehouse operations and how they can be achieved in a safe manner. Common belief may suggest that a slower forklift is a safer forklift. However, with wider aisles safety can be achieved while improving productivity.

It is clear that warehouse working and traffic conditions may change daily or even hourly. The list of varying previously mentioned factors, although not exhaustive, illustrates the importance for receiving personnel to check freight closely for damage or other freight conditions that may cause hazards when checking, moving, or putting away product. Workers must rely on all their physical senses to identify any potential issues in the work place, such as seeing that visitors are present who may not respond in a predictable manner when around unfamiliar warehouse equipment and freight, and listening for unfamiliar sounds that may indicate falling objects or contact between freight, equipment, and storage systems. The Bureau of Labor Statistics reports that for all workplace occupations, the total fatalities due to employees being struck by objects or equipment have increased by 7 percent from 2011 to 2012. As short-story author Robert Fulghum wrote, "All I really need to know I learned in kindergarten." Although not as simple as that, utilizing four of the five senses (less taste) can help warehouse employees create a safer workplace for themselves and the customers' product.

From the highly sophisticated and automated-assisted FedEx document sortation and transport facility to the frontline materials handling and assembly line of Hyundai automobiles, human senses are leveraged to improve efficiencies and safety. For example, the use of a Kanban visual signaling process alerts materials replenishment operators that more of a particular SKU is required.

Materials inventory positioned at frontline assembly points is reduced because visual cues with Kanban cards or electronic lights indicate the need to replenish so that the material arrives "just-in-time" of need. Similar to the benefits of an active and reserve inventory layout, a Kanban-type process, when employed, helps to keep a majority of the workforce away from a majority of the inventory.

## Safety Cues

Electronic information boards provide workers and supervision critical information for production, as well as, alerting employees when help is needed at a specific point in production. Information boards such as these are positioned throughout the materials handling and assembly processes within auto assembly plants, such as the FedEx package and document sortation facilities in Memphis, TN, and the John Deere tractor assembly plant in Mannheim, Germany. Moreover, some of the plants employ audio and visual cues that can be detected plant-wide to alert employees of important information. In one facility, employees work within assigned area groups. Each group selects a particular song melody that plays when the need arises to alert employees and management of a problem experienced by the work group. When the song is broadcast, everyone knows exactly where in the production process the problem lies. Sensory-based cues also are utilized to assist in the safe flow of materials from supply to staging parts at the frontline of production.

## Visual Types of Cues

Visual types of cues are indicators for the senses that pertain to seeing clearly within the confines of the warehouse space a worker occupies. Visitors and new employees, unfamiliar with the working warehouse, may inadvertently walk into freight movement traffic. Forklift operators must always be aware and perceptive while operating a moving lift. Workers with their backs to you or heads turned away may not know you are there or may respond in an unpredictable manner. Varying work takes place in a warehouse. Clerical duties, including checking freight or performing inventory cycle counts, may require the attention and focus of clerks. They may be focusing on product documentation instead of the vehicles moving around them. Product out of place may cause accidents when unexpectedly it is in the forklift operator's path or when trying to move around it. New types of products that may sit differently on a pallet by overhanging or creating a high center of gravity for a palletized load present an unsafe situation. Equipment operators unfamiliar with handling oversized palletized product may have difficulty when moving product around corners beside other freight or situating the product within its designated slot location.

Other visual hazard cues include but are not limited to

- Dunnage on the floor may tangle in forklift wheels, become airborne when run over, or become a tripping hazard.

- Leaning palletized freight may fall.

- Fluid on the floor may cause slipping or indicate a mechanical issue with the fork lift.

- Leaking packaging may cause contamination of other products, slipping hazards if it leaks on the floor, or may infect the integrity of the packaging causing the freight to topple.

## Audio Types of Cues

Audio types of cues are indicators for the senses that pertain to hearing clearly within the confines of the warehouse space a worker occupies. Unfamiliar sounds may indicate the presence of unaware visitors or personnel. When equipment operators perceive to hear voices they do not recognize, it becomes important for safety that they work to see the pedestrians and ensure that they are not nearing the operating machinery. Counterintuitive perhaps, but a lack of expected sounds may indicate that product is not exactly where expected. For example, when a pallet is lifted and set into place in a rack, the operator may equate the proper rest of the pallet within the racking slot location. Absence of the audio cue may signify that the pallet of product has not yet reached the resting place within the slot. An attempt to remove the fork attachment may inadvertently unsettle the pallet on the lift and rack causing a hazard. Hissing noises originating from a product canister, tank, or drum may indicate a hazardous breach in the wall of the packaging, and thus, allowing product to escape. Forks on lifts come into contact with freight on a regular basis. Although unexpected, the contact may cause the forks to tear into the industrial packaging designed to protect the product. Lift operators must be vigilant in recognizing such damage and notify supervision of the accident before it becomes an escalating hazard.

Other audio hazard cues include but are not limited to

- Unexpected loud machinery could signal a mechanical issue or contact between machine and product or storage structure.
- Unexpected loud voices may indicate an emergency.
- Startling noises may indicate falling freight.
- Horns indicate presence of other vehicles.
- Sirens, buzzers, or whistles may indicate an emergency.

## Touch Types of Cues

These are indicators for the senses that pertain to feeling or touching product within the confines of the warehouse space a worker occupies. Packaging that is wet to the touch may signal a product leakage. *Leakers,* as they are called in the warehouse, have the potential to damage product below them through contamination and by damaging the industrial and marketing packaging of the product. Leakers are known for softening the integrity of corrugated cardboard packaging and thereby causing palletized product to collapse and shift unexpectedly.

Other touch types of hazard cues include but are not limited to

- Slippery floors may indicate leakage in product packaging or machinery.
- Resistance when moving equipment may indicate an unexpected obstacle.
- Unexpected bump when driving equipment may indicate running over an obstacle.

## Odor Types of Cues

Other types of cues are indicators for the senses that pertain to the smelling of unfamiliar odors existing within the confines of the warehouse space a worker occupies. Unfamiliar odors may indicate the presence of gas or smoke that is indicative of explosive or fire hazards. Odors that should not be present at the same time or place may indicate that two products are stored together that should be stored apart. Contamination of porous products may take on the odors of other products within close proximity. Manufacturers should provide warehouse operators directions for storage of such products so that one is not contaminated by another. Regardless, warehouse materials handlers, clerks, general workers, and supervisors must be aware of and investigate unfamiliar odors. Although perhaps not of immediate danger to personnel, a repugnant odor may indicate rodent issues or even infestation.

Although the list of cues may seem endless for recognizing and identifying warehouse hazards, by training personnel to leverage their senses and awareness of warehouse environmental anomalies, the workplace can become safer for people and product. Accidents may still occur, but training and awareness can help to reduce the threat and occurrence of accidents.

## Safe Picking and Replenishment

Chapter 4 discusses product layout. The strategy of creating a reserve and active picking process helps warehouse operators to position inventory and workers where the greatest interaction takes place in the smallest area. For filling orders and loading trailers, an active area situated near the shipping dock facilitates the efficient movement of products for shipping. Safe work areas are maintained by keeping a majority of the workforce out of the reserve stocking areas and within the active picking and shipping area. Replenishment operators should be the only product handlers moving product from reserve to the active picking area. Training specific personnel to perform replenishment duties helps to maintain safety of workers and products. Order pickers routinely assigned to a specific area become familiar with product location and their sensory cues should be heightened when something doesn't look, sound, feel, or smell normal or as expected. This should also be true for employees assigned to zone picking.

The environment becomes more complex when automated assisted machinery is employed in the same location as humans work. However, sophisticated automated-guided vehicles (AGVs) are equipped with line-of-sight laser sensors to navigate the AGV and product safely through the warehouse. Safety sensors scan for obstacles in the AGV's pathway. Technology stops the AGV when an object unsafely enters into the pathway. Of similar concern is the co-location of personnel and conveyors, carousels, and other automated product moving and storage equipment. Conveyors are equipped with sensors to help identify when personnel are in jeopardy.

Warehouse management systems create efficiency gains in some operating environments by directing forklift operators to their next move nearest the operator. Directives are established as operators complete and confirm their assigned product movement. Sophisticated algorithms take into account the location of other employees and equipment and the work performed, as well as, the remaining work tasks to be accomplished. Forklift operators are then assigned the next most-efficient move for the warehouse operation.

*Task interleaving* helps to reduce operator travel time and enables the efficient utilization of lift operators for various work tasks. One move may require an equipment operator to put away a received pallet of product into the reserve section of the warehouse. By electronically verifying within the computer system that the move has been accomplished, the operator is provided another task originating near her current position in the warehouse. The next task may require the operator to pull from storage a distressed product pallet and transport it to OS/D for rework and repackaging into knockdown cartons.

A potential concern of task interleaving is that a forklift operator must become familiar with the sections of the warehouse in which the various tasks must take place. Task interleaving by zone may help to obtain equipment efficiencies, while ensuring that the lift operator maintains inventory integrity, safety, and productivity. Upon instituting task interleaving to assign forklift personnel duties, one distribution center realized a drop in productivity for unloading trailers. When asked, the lift operators explained that they no longer could evaluate their individual productivity by trailer unloads for the shift. As it turned out, receiving lift operators had a healthy ongoing competition to see who could unload the most trailers per shift. Task interleaving prohibited the competition. When managers for the DC realized this, they absolved forklift operators unloading trailers from task interleaving. The remaining materials and product handlers remained performing their task interleaving duties.

## Safe Product Staging

Whether staging an inbound shipment or outbound shipment, the loading dock can be ever-changing with product types and quantities. To complicate the safe management of the loading dock, often drivers for carriers are required to inspect and count the freight they receive. Over the road drivers may be unfamiliar with the facility and loading dock process, and safety may be an issue. Number of loads being staged may also vary, and so may the accompanying drivers, forklift operators, and receiving/shipping clerks.

Several strategies exist to reduce the interaction of personnel with freight, such as warehouse personnel or carrier personnel. For example, institute a shipper load and count (SLC) program, whereby, the warehouse is responsible to count and load the freight without requiring outbound carriers to check the freight. Not only does this simplify the loading process, but it also contributes to a safer shipping loading dock because transportation personnel need not be present on the shipping dock during the freight loading process. Perhaps a better strategy is to work with carriers, shippers, and the shipping department of the warehouse to design a safe and secure drop trailer program. Safety is enhanced because trailer loading can be accomplished during lighter workload periods in which fewer lifts are in operation and fewer carriers are present to count and live-load freight. When loaded, carriers can schedule trailer pickup during off-peak hours so that safety and efficiency is maximized for drivers and warehouse operators.

The application of RFID tag technology for product handling within the warehouse has been shown to be good for productivity gains and improved safety. Improved safety is achieved when product no longer has to be counted by checkers, lift operators, or truck drivers. RFID tags are read as palletized product moves through an RFID tag reader corridor consisting of antenna activating tags and recording data pertaining to product movement into the trailer. Tags and the accompanying readers perform the product checking duties and alleviate the need for clerks or carriers within the loading area. Safety is enhanced.

Staging palletized products in the sequence to be loaded onto a trailer reduces the need for a lift operator to shuffle pallets around to locate the next in line for loading. The effective staging area for a global brand apparel marketer includes building loads within staging as they will appear within the trailer. Lastly, a simplified process would seek to remove the delay or dwell time associated with staging product. Refined shipping processes would allow forklift operators to travel direct from storage slot location into the trailer. RF technology affixed to forklifts or the advanced RFID tagging technology may be utilized to achieve such efficiencies and safety. Research on RFID technology gains within the warehouse supports the correlations between the technology and employee safety.

## Securing the Product

By nature, warehouse environments bring together people, product, and machinery. Inside employees and outside service providers often work together within confined and varying spaces. For this reason, it is critical to understand the factors impacting the safety and security of product within the workplace.

### Pilferage and Theft

Safety of the product includes protection from damage and theft. From the minor lifting of a candy bar for a snack from a carton in storage (informally referred to as *grazing* in packaged food distribution) to the more sophisticated planning and hijacking of a trailer loaded with expensive electronics, theft in the warehouse, although not a particularly positive subject, must be acknowledged and a workplace culture fostered that aims to protect the company's and employees' financial and job interests.

Not to make light of the subject, it reminds me of the song, "One Piece at a Time," sung by Johnny Cash. As the story in the song goes, while working on an auto assembly line in Detroit from 1949 through the 1960s he took home all the car parts, "one piece at a time," needed to eventually build his own Cadillac. How did he sneak out the parts? Some were hidden in his lunchbox, while bigger pieces required the assistance of his buddy's vehicle. Although fictitious, the folksy song suggests some level of reality, in that, employee theft may occur in many forms and by one or multiple individuals working together. Here was another problem: In the song, he didn't consider it stealing nor did he consider himself a thief; and, according to the character portrayed in the song, the company wouldn't miss one little car part if it went missing only every so often.

A problem existed within one privately managed distribution center dedicated to storing and moving products to general product retail stores in the southeast (based on one author's discussion with a manager in the distribution center). Employee pilferage within the facility became such a costly problem for the company that security guards were positioned at the entry and exit point of the building. When reporting for work and at the end of the working shift, employees were subject to search and screening through a metal detector.

Needless to say, the working culture within the distribution center had to be changed. The immediate problem was solved with the search and screening procedures. However, the long-term solution required management to help employees realize how theft, no matter how small and seemingly insignificant, actually correlates with levels of pay increases and potential decreases workforce-wide. Management erected a visual display showing a bar chart of the dollar amount associated with internal theft for the facility. Employees were shocked at the cumulative dollar amount. In Johnny Cash's song, the theft of "one piece at a time" in small amounts over time equated to a $100,000 cumulative theft in the end.

In the actual distribution center, management then made the appeal to the employees that the dollars lost to internal theft of product "walking out the door" caused an equal loss in employee raises. The employees got the picture. Dramatic reductions in theft within the facility occurred, the screening and detector equipment was eventually removed, and employees saw eventual raises in pay. The culture was changed through discussion, education, and setting expectations. The efforts even resulted in significant reductions in employees taking, without permission, another employee's lunch stored within the break room. Yes, the company would terminate the employment of a worker for even the smallest of theft of property belonging to a co-worker. Employees began looking out for the welfare of each other and for the company. The entire facility began a campaign of zero tolerance to employee theft.

Countless stories exist about theft of products being stored throughout the various nodes and links constituting supply chains. Stories include organized theft of cigarettes yet to be taxed to stories of a carrier knowingly receiving freight unintended for his pickup (that is, product overage). Carriers receiving overages but keeping them and deliberately failing to report the additional unrecorded freight shipped are known as "making" cases of product. The deception practice may escalate when adversarial relationships exist between carriers and shippers, and when carriers feel unfairly or overwhelmingly burdened with freight claims from shippers. One of the authors visited a carrier within his former company's network only to recognize that product cartons belonging to one of the warehouse's top clients were stored in the back of the carrier's cross-dock. The carrier was storing product overages instead of returning them to the public warehouse. In the case that the carrier received a claim for lost product, the carrier would return the same amount of overages and negotiate with the shipper or warehouse to allow the overages to offset the lost cases of a completely different SKU.

Although it may work out that similar items were inadvertently picked and loaded resulting in an overage of one SKU and a shortage of another. Oftentimes, customers allow substitute products if the products are similar in use and application. But the carrier was in the wrong by storing unknown overages only to protect the carrier from future product shortages.

The category of product theft can and should be broadly defined. Following is a brief list of some of the situations that may arise:

- Warehouse worker collaborates with an outside truck driver to leave freight on an inbound trailer but indicate on the bill of lading and delivery receipt that the freight was received.

- Product cartons left in dock doorways transfer to an employee's car in the parking lot.

- Product cartons are removed from the warehouse, hidden in the brush or behind trailers, railcars, or other visual obstacles, such as garbage bins or recycling compactors. Hidden product is the removed after operating.

- Knowingly substituting a more valuable SKU for a lesser valued SKU, but not documenting product substitution on the picking, checking, loading, or shipping documents.

Although countless stories may exist of theft within warehouses and distribution centers, perhaps a few core steps may be taken to avert would-be-theft. The steps require effort on the part of management, employees within the warehouse, and carriers entering and interacting with warehouse employees and product.

First, like the distribution center that changed the culture within its workplace, managers and employees must see that both are exercising high character and have no tolerance for deception and thievery. Moral, ethical, and legal workplace behaviors must be fostered and expected at all levels and with any persons entering the facility structure or grounds. Although research exists to indicate the importance of managing warehouse operating environments and personnel, of the research pertaining to personnel issues published within the leading logistics journals, Daugherty et al. (2000) found that senior supply chain professionals believe that the managers reporting to them exercise high-integrity attributes. However, none of the research in the journals pertained directly to measures to take in reducing theft by employees within frontline logistics operations. The body of research pertaining to logistics personnel and published within leading logistics journals is outlined in Keller and Ozment (2009).

Single entry and exit doors should be required. Employees should understand that all other doorways are to be secured (according to fire codes) and only utilized for emergencies. Dock doors should be open only during trailer occupancy, and after a trailer pulls away from the dock, the door should be lowered and the appropriate stop/go lights illuminated. A clean area contributes to a safer and more secure workplace. Designated and restricted areas should be dedicated to trash and recycle bins. Although it may be difficult to utilize, bins should be located away from entryways so that product cannot easily be dumped into a bin, only later to be retrieved and removed from the facility. Similarly, vehicles other than delivering trailers should be located away from warehouse entryways and exits—even for emergencies. Safe pedestrian walkways should indicate safe walk path and cross walks for employees and visitors to move from parked automobiles into the facility.

Product rework zones should be off limits to personnel not assigned to the areas. Exposed product often dwells within OS/D areas and provides temptation for pilfering. Actually, damaged freight awaiting disposition or rework should be placed in out-of-the-way locations or slots that are inefficient for routing product putaway and picking. Mezzanines over doorways offer space that is unattractive for other duties and may be appropriate for storing product requiring rework.

Staged transportation equipment should be well lighted and trailer doors closed when not positioned against the dock. Employees were caught removing cases of untaxed cigarettes and hiding them behind and inside an open railroad boxcar staged on a rail siding in the rear of the warehouse.

Of equal concern is the theft of cargo while sitting in the terminal yard of a warehouse. Documentation should indicate the identifying trailer or container seal number and condition of the seal upon receiving freight into the terminal. An in-tact seal notated on the delivery receipt and copy of the bill of lading should indicate the trouble did not stem from the delivering carrier, and that theft occurred within the warehousing compound.

## Damage

Primary opportunities for damaging product include shifting loads during transit, handling product during unloading, any stopping-dropping-picking up points, and when placing in storage in stacks or within racks. Moving products confined to tight spacing with other products often results in one pallet connecting with another. During the move the unintended pallet may be dragged along with the intended pallet causing both the driver and product to be in jeopardy.

Lifting product pallets to heights above the second pallet slotting position requires operators skilled in visually identifying when the product clears the bottom rail of the slot location. Optical technology exists within the warehouse allowing lifts to utilize optical scanners to identify when the palletized freight clears the slot opening and it is safe to place the freight forward into the slot. As shown in Figure 13-3, programed automated storage and retrieval systems allow for routine and constant accuracy to be achieved by leveraging the capabilities of scanners and programmed lifting equipment.

Figure 13-3    Automated storage and retrieval system

Superior quality and process control within the warehouse requires an organized workplace. Out-of-place product causes hazards to product and personnel. Having a well-organized product flow and storage strategy and an overall facility layout helps to retain familiar product positionings and reduces products in aisleways and intersecting paths.

Damage may also be caused by the handling equipment utilized to move the product. For example, one loading dock transit shed operator was responsible for unloading large rolls of imported paper destined for printing the *Los Angeles Times* newspaper. Not enough vacuum suction cup devices were available for the productivity planned for the shift, so the lift operator decided to utilize clamps placed on his lift to squeeze and lift the paper rolls to move them to and within the transit warehouse. Unfortunately, the equipment operator was not adequately skilled, and consequently applied too much pressure while clamping the material causing a 5-inch gash in the paper. The entire 5 inches was damaged beyond use and represented an enormous amount of paper material because it was so thin and tightly wound on the spool. Damage costs ranged in the thousands of dollars. Clamp devises have been known to crush light product packaged within standard-type corrugated cartons. Although push-pull clamping devises grab onto slip-sheets and pull pallet quantities of light product onto the forks, they often dig into the bottom cartons and damage product.

Equipment operator training and skill development helps to protect the product from operator error. Moreover, it is important to select the appropriate materials handling equipment by considering the handling characteristics of the product.

## Fire and Water

Storage locations may expose product to conditions in which they may be vulnerable to fire and water, and extra caution may be a legal requirement. There are oxygen cylinders that are sensitive to high-heat exposure that may be emitted from a high-wattage light bulb. Managers must well understand the products' sensitive characteristics prior to storing product within the facility.

In one incident, a seasoned mechanic inadvertently left the water faucet turned on over a 3-day holiday weekend. A hose was attached to the faucet, but the spray nozzle attached to the hose was in the off position, so the water appeared to be turned off. Pressure from the water caused the hose to burst, and water began filling the warehouse floor and continued for 3 days. Sugar substitute product was warehoused in the facility, and the corrugated cartons absorbed the water and began to collapse under the pressure of the higher-stacked cartons.

Low-lying flood zoned locations may prove to be hazardous to storing product. For this reason, potential tenants or builders must retain documentation pertaining to the zoning and history of rising and flooding water in the location. Locating near river levees have proven risky in recent years with overflowing rivers caused by flood rains and hurricanes.

## Infestation

Another potential threat to some products comes in the form of infestation by insects. Imported forest products must be inspected for wood-devouring beetles that if entered into an area could prove devastating to the areas forest. Pallets containing imported products have been known to harbor harmful insects. Quarantining the pallets will be the first step to stop the potential spread of harmful insects. Fumigation steps and destruction of the source is of great importance. It is up to warehouse operators and inspectors to detect the presence of infested product or packaging by understanding the signs of infestation and removing samples for testing.

# Physical Security Measures

Safety and security of product and personnel often requires separation through physical means. Although the U.S. Customs regulations do not require interior fencing to isolate imported product yet to be cleared by Customs, the threat of not providing a gated fence area within the warehouse to secure the product will ultimately be the responsibility of the warehouse operator. For example, while in the final stages of applying for activation of a Foreign Trade Zone (FTZ), a Customs agent highly recommended the construction of such a fence. Such a fence would limit the space utilization of the public warehouse for storing other client products when the space was not occupied by imported product yet to be Customer cleared. An appeal to the regional Customs Office resulted in approval for the warehouse to elect whether to build a fence to secure the imported freight. However, the letter from the regional office clearly stated that any loss or inadvertent shipment of imported and uncleared product from the FTZ would result in a financial penalty attached to the warehouse bond required to receive activation authority.

Secured fencing is common for segregating high-valued product from other freight within the warehouse. For the untaxed cigarettes, they too were placed within a secured fenced and locked section of the warehouse. Even with added walled security, theft can occur. Other physical security measures may include automatic opening and closing doorways that create firewall barriers between warehouse sections. Alarm systems alerting management that an emergency exit has been accessed is another security measure that can be taken.

Although no discussion can be comprehensive enough to cover all safety and security issues and measures important for operating a warehouse, this chapter illustrates potential hazards including, but not limited to, theft, damage, injury, infestation, fire, water, and flooding. Most important, it is incumbent on warehouse management and supervision to instill a commitment among all employees and create an on-going culture that understands the negative impact of safety violations and security breaches, and a workforce dedicated to protect the workplace. In one Oregon coastal warehouse supporting the manufacturing of consumer paper products, management stepped aside to allow the employees to conduct start of shift safety meetings. When a safety issue is brought up, the workers assign each other duties to research the issue and come up with viable solutions to make the workplace safer. The next meeting the assigned employees report their findings. Fostering a workplace whereby co-workers are accountable to each other and to supervision as well as supervision being accountable to each employee can help to ensure safety and security in the workplace. Research supports this proposition.

Supply chain security is ever-increasing in importance. Research indicates that businesses are of greatest exposure to security threats. Links and nodes within supply chains can be used as targeting points for attacks, or logistics assets could be used as a weapons' delivery systems (Voss, Closs, Calantone, Helferich, and Speier 2009; Voss 2006). Warehouse security costs have been shown to reach $2 per square foot. And as Figure 13-4 indicates, there are challenges between managing business priorities and managing security priorities.

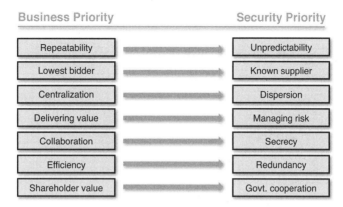

Figure 13-4    Conflict between business and security priorities
(Courtesy of Doug Voss 2006, p. 25; see also, Sheffi 2001)

There are questions that still remain about who is willing to pay for the cost of security and when a supply chain partner is most likely to step forward. Price and reliability compete heavily for resources that are also in need for increasing security in the supply chain. Those firms having survived a security breach are more willing to make the trade-off and invest in greater security measures so as not to repeat the security failure.

## Summary of Key Points

Safety of personnel and security of product are the basic building blocks for warehouse operations. Neither can be neglected without detriment to operational efficiency and the bottom line. The warehouse manager must set the tone in these two areas and work to control each on a daily basis. Emplacing the proper procedures, executing constant vigilance, and continuously training personnel to standards assists in maintaining good order and discipline when it comes to safety and security. Securing products, infrastructure, equipment/personnel resources and the public is an important issue for all supply chain partners. Who in the supply chain will bear the costs to increase security continues to be debatable. None of this can be left to chance, however. The steps discussed in this chapter must be actively and continuously applied to ensure a safe and secure workplace.

# Key Terms

- Active Product/Picking Area
- Area Picking
- Automated Guided Vehicle (AGV)
- Automated Storage and Retrieval Systems (AS/RS)
- Bill of Lading
- Bone Yard
- Carousel Picking System
- Containment Storage Rooms
- Distressed Product
- Distribution Center (DC)
- Dock Bumpers
- Dock levelers
- Drayage Companies
- Drop Trailer Process
- Dunnage
- Foreign Trade Zone (FTZ)
- Grazing
- Infestation
- Inventory Flow
- Inventory Integrity
- Inventory Turns
- Inventory Velocity
- Kanban System
- Knockdown Cartons
- Leakers
- Longshoreman

- Material Handling Equipment (MHE)
- Motor Carrier
- North American Industry Classification System (NAICS)
- Order Cycle Time
- Over the Road Driver
- Overages, Shortages, and Damaged (OS/D)
- Picking
- Pilferage
- Product Staging
- Radio Frequency Identification (RFID) tags
- Receiving Clerk
- Red/Green Light Trees Replenishment Operators
- Reserve Inventory/Storage Area
- Sensory-based Cues
- SKU
- Shipper Load and Count (SLC) program
- Single-Double Lifts
- Staged Cargo
- Super Cargo
- Task Interleaving
- Transit Shed
- Utility Tractor Driver (UTR)
- Warehouse Management System (WMS)
- Wheel Blocks
- Zone Picking

# Suggested Readings

Akerman, K. B. (1997, 2012), *Practical Handbook of Warehousing*, 4th ed., Chapter 22, Chapman and Hall, New York, NY.

Daugherty, P.J., R. F. Lusch, M. B. Myers, and D. A. Griffith, (2000), "Linking Compensation and Retention," *Supply Chain Management Review*, Vol. 4 No. 4, pp. 64–72.

He, Y., J. Wang, Z. Wu, L. Hu, Y. Xiong, and W. Fan,. (2002), "Smoke Venting and Fire Safety in an Industrial Warehouse," *Fire Safety Journal*, Vol. 37, No. 2: 191–215.

Keller, S.B. and J. Ozment, (2009), "Research on Personnel Issues Published in Leading Logistics Journals," *International Journal of Logistics Management*, Vol. 20 No. 3, pp. 387–407.

Martens, B. J., M. R. Crum, R. F. Poist, (2011), "Examining Antecedents to Supply Chain Security Effectiveness: An Exploratory Study," *Journal of Business Logistics*, Vol. 32, No. 2: 153–166.

Sheffi, Y., (2001), "Supply Chain Management Under the Threat of International Terrorism," *International Journal of Logistics Management*, Vol. 12, No. 2, pp. 1-11.

Swartz, G. (1999), *Warehouse Safety: A Practical Guide to Preventing Warehouse Incidents and Injuries,* The Rowman and Littlefield Publishing Group, Inc., Lanham, MD.

Tompkins, J. A. and J. D. Smith, (1998), *The Warehouse Management Handbook*, Tompkins Press, Raleigh, NC.

Voss, M. D., (2006), *The Role of Security in the Supplier Selection Decision*, Dissertation, Michigan State University.

Voss, M. D., D. J. Closs, R. Calantone, O. K. Helferich, and C. Speier, (2009), "The Role of Security in the Food Supplier Selection Decision," *Journal of Business Logistics*, Vol. 30, No. 1, pp. 127–155.

Williams, Z., J. E. Lueg, S. P. Goffnett, S. A. LeMay, and Cook, R. L. (2012), "Understanding Supply Chain Security Strategy," *Journal of Transportation Management*, Vol. 21, No. 1: 7–25.

Williams, Z., N. Ponder, and C. W. Autry, (2009), "Supply Chain Security Culture: Measure Development and Validation," *International Journal of Logistics Management*, Vol. 20, No. 2: 243–260.

# 14

# EQUIPMENT AND INFORMATION TECHNOLOGY

## Introduction

By employing a warehouse management system (WMS), a transportation management system (TMS), and other information management tools now available on the market, managers can better respond to changing conditions by leveraging the data accuracy and computing power inherent in today's integrated information systems. In doing so, better efficiency and reductions in cost can also be realized. Operational goals and corporate goals must be well defined and understood before selecting a WMS. Specific warehouse needs and customer requirements must also, first, be well defined. This chapter discusses the advantages that a WMS provides and presents an analysis toll for selecting the most appropriate WMS. This chapter also covers additional technology enablers and various types of equipment to use in support of warehouse operations. The benefits of employing the appropriate technology are also expounded upon.

It is not likely that handling and storage equipment discussions can persist without equal discussion on the application and utilization of technology in the warehouse, such as information based or mechanization technology. For example, the simple process of applying stretch wrap or the transparent film wrap around palletized cartons to bind the cases into a single unit can now be accomplished manually or mechanically. Industrial-sized bags of plastic-like pellets may be sealed, palletized, and stretch-wrapped at the end of the production line and without human interaction. Downstream the supply chain the same material, but now processed into finished automobile parts, may also be mechanically stretch-wrapped for protection during international transport to the production line at the final auto assembly plant. The automated and mechanized palletizing and stretch-wrapping can help to reduce costs associated with labor, packaging materials, and damaged cartons due to uneven wrapping.

Advances in machine technology have proliferated from the production line to the warehouse and distribution center. Machine technology and information technology in the warehouse enable internal integration across traditional functional departments and external integration with suppliers and customers.

## Warehouse Management Systems (WMS)

One author recalls learning to create consolidated freight loads so that the warehouse company could offer transportation savings to its clients. Shipping alone, an individual client order may have only enough volume to transport under less-than-truckload (LTL) rates. Consolidating the orders of multiple warehouse clients, however, allowed the warehouse to gain the economies needed to ship via truckload carrier, which charged a lower rate per 100 pounds (hundred weight or cwt.) compared to the LTL cost for individual shipments. The more accurately the warehouse identified and scheduled consolidated orders, the better the profit margins it would receive for its consolidation services.

Scheduling daily consolidations for the warehouse could require hours of tedious work to evaluate the consolidating factors of each order for the following:

- Required shipping date
- Required delivery date
- Destination ZIP code zone
- Freight classification (associated with the ease or difficulty and cost to handle the specific SKU)
- Weight
- Cubic dimensions
- Hazardous warnings
- Client instructions
- Any special instruction of the consignor or receiving customer

The objective was to create as many multiple order consolidations associated with the greatest possible total cost efficiency and accuracy. It was clear that continuing to perform the necessary optimization calculations by hand would not allow the warehouse to be competitive in the future. Three months were required for the scheduler and programmer to create an expert system to conduct the consolidation decisions.

Technology capabilities today enable companies to perform such optimization analyses much quicker than when performed by hand. In the previous example, the manual system could require from 1 to 3 hours to complete. Accuracy and timing were dependent on the skills of the manual scheduler and the rate at which orders entered into the system. With the help of today's execution systems (for example, Figure 14-1 illustrates a warehouse management system) and corporate-wide enterprise resource planning (ERP) systems, managers are equipped to conduct multiple what-if scenarios pertaining to the daily, weekly, and monthly operating requirements, resources, and constraints. As customer, carrier, resource, or process conditions change, planners and managers can better respond to new conditions by leveraging the data accuracy and computing power inherent in today's integrated information systems.

Figure 14-1    Technology utilization in the warehouse

Information management systems for guiding decisions and operations within the warehouse may be a corporate developed program, an application offered through the company's ERP system, or a standalone system that integrates information exchanges with the existing ERP system of the company. An alternative to the traditional on-site option, there are now web-based WMS applications complete with cloud-based computing and storage.

Certainly, total cost of owning or utilizing sophisticated integrated WMS software is a large consideration when evaluating and selecting a vendor and product. Of equal importance, however, is the need to identify the warehouse or distribution center's (DC) needs so that the most appropriate WMS system can be selected to provide the most appropriate functionality. In addition, while standalone labor management systems and terminal yard management systems exist, more of today's leading WMS products offer these functions within their WMS software.

Costs that must be factored into the WMS purchase decision include, but may not be limited to, necessary hardware and networking machine costs, software and user licensing costs, user education and training costs (also the number of users may influence the overall pricing), implementation costs, technical support costs, and costs associated with current and future customization needs. Of course, these costs pertain to those associated with the WMS vendor. The company must also quantify the relevant internal costs associated with personnel cost to research alternatives, employee and supervision costs during training, and learning curve costs associated with the start-up implementation of the new system software.

Corporate and operational goals must be well understood before selecting a WMS. Customer needs and warehouse operation needs must also be well defined. Processes and the activities making up the processes must be taken into consideration during vendor evaluation. Critical information such as this helps to determine the warehouse functions that are most important for achieving company and customer warehousing goals. A WMS offers varying scopes of functionality pertaining to the major processes central to managing warehouse operations. From utilizing advanced shipment notifications (ASNs) to assigning receiving doors, a WMS helps to remove the guesswork when making operational and tactical decisions that aim to improve process efficiencies and resource utilization. Both should lead to quality improvements and cost reductions. Receiving, replenishment, picking, shipping, inventory management, labor management, transportation management, and other functions allow for an efficient order fulfillment process.

## Functionality in Receiving

Receiving processes rely on inbound ASNs so that advanced labor and equipment scheduling can be assessed. The process continues with providing carrier appointments based on labor, equipment, and dock doors availability. A WMS must accommodate documentation verification and optimal door assignment based on load contents, SKU stocking locations, orders, and in-stock availability to determine cross-docking opportunities. Material handlers can receive immediate and optimal slot locations for putting away received units. Functionality within the WMS includes providing customer-driven standard decision rules for determining product disposition of returned product. Dropped trailers (loaded or empty) are easily identified and dispatched for exit or worked into the unloading and receiving sequence as labor, equipment, and dock doors allow. Radio frequency identification (RFID) technology has been shown to significantly improve efficiencies during the receiving process. This is discussed in the "Radio Frequency Identification (RFID) Tag and Barcode Technology" section later in the chapter. Integrating

RFID and WMS can equip managers with the best information and knowledge to make the lowest cost and most effective decisions.

## Functionality in Replenishment, Picking, and Shipping

WMS systems take into account replenishment levels designed to increase inventory velocity. Reserve locations are maintained and active picking locations are monitored. Operators receive dynamic instructions for the next in-line SKU slotting assignment. In this manner, the WMS helps ensure accurate product rotation and efficient item placement for the picking process.

Varying products influence stock layout within the warehouse. Moreover, order requirements and SKU characteristics are evaluated to determine the most appropriate and cost-effective combined picking and shipping strategy.

Capacity availability in space, equipment, and personnel may make wave picking the best strategy for a warehouse. The WMS helps to optimize and schedule pick waves, manage zone picking strategies through employee assignment, and generates shipping documents for advanced notice to customers and for including in the shipping carton.

## Functionality in Managing the Execution of Transportation

To route an order or shipment, managers must rely on their knowledge of the customer and product, shipment characteristics, such as weight, cube, destination, ship and delivery dates, and other information critical for shipping and delivering the freight on time. WMS systems may come with a routing function or may integrate with a transportation management system (TMS) to facilitate the shipping process. Transportation management requires optimizing mode and carrier selection, along with optimizing load building configurations and even building consolidated loads from multiple individual orders.

Features of a WMS transportation management function also may include assigning carriers based on most favored freight rate to transport orders, but also weighted by the service level quality of the carrier pool. Freight costs are calculated with utilizing a rating function, and the auditing of freight bills, invoice for requesting payment, occurs by the WMS comparing the shipment data and tariff (agreed pricing) on file with the request for payment information detailed on the freight bill. Claims management is made easier with electronic availability of original document information, logging of authorization, and receipt signatures along with the transit and certain pairwise documentation auditing capabilities.

## Functionality in Order Fulfillment and Inventory Management

From digital order transactions to facilitate order fulfillment to cycle counting and managing annual physical inventories, the WMS can manage the many needs associated with complex and dynamic inventory and order processing. Credit checks and checks for the availability of inventory are efficiently accomplished through the WMS application.

## Functionality in Managing Labor

Variables influencing labor decision for the warehouse assignment may include hours available considering weekends, holidays, and scheduled vacations. In addition, employee skills and productivity levels are captured within the WMS. Some machine operators may simply perform one task, such as trailer unloading, whereas another thrives on varying assignments through task interleaving. Managers may work within the WMS to assign appropriate work to the appropriate worker. Daily workloads may be more equitability distributed on light and heavy workload days. One progressive DC placed information kiosks in break areas for employees to access their employee accounts to schedule vacation time, volunteer for extra work, schedule training, and even submit suggestions or complaints. The kiosk also provided employees with current news about the company and employee benefits.

### WMS Vendor Selection

It is clear that the better the fit of the WMS to the specific warehouse needs, the better the product will flow through the facility with lower discrepancies and with lower costs. Of course, WMS programs don't come free. There are costs associated with the time it takes to research potential WMS providers and the software itself. A Request for Information (RFI) would be a good starting point after preliminary sources are identified. Information will likely entail an on-line or in-person demonstration of the vendor's product. Prior to that, however, the manager must seek internal information from employees and managers to identify the specific functions of importance to the warehouse operation. By providing potential vendors with this information, they will be better equipped to explain and illustrate how their software will fit the warehouse's specific needs. A weighted scoring process can compare vendors on the most important criteria and determine a ranking of the top potential vendors.

It is highly likely that a cross-functional team may be placed in charge of evaluating the vendors and recommending to senior management a ranking of the top two or three. Analytical hierarchy processing (AHP) is a tool that utilizes pairwise comparisons between selection criteria and then between vendor candidates/WMS software on each of the selection criteria. The results of the analysis allow the evaluation team to conclude the following:

- A ranking of the highest preferred criteria down to the lowest preferred criteria

- A ranking of the vendor scoring highest to lowest on each individual criteria

- A final ranking of the vendor scoring overall highest to lowest when considering all the criteria and vendors

AHP allows for the comparison of multiple suppliers based on multiple selection criteria. It allows for both subjective and objective data analysis. It is particularly useful in attempting to quantify multiple qualitative variables and facilitates group decision making.

You perform these kinds of analyses in your daily personal and professional life. A graduating student faced with a decision to select one company's job offer over three other job offers will begin tedious contemplation of the job criteria most important to her. Salary, other benefits, location, the work itself, reputation of the company, ability to advance, vacation and holiday time, and other factors likely will be considered. The task of selecting between the four job offers becomes more complex when each offer varies from the others across the selection criteria. The comparisons can become frustrating. By writing down the criteria and companies and processing the comparisons through this weighted score approach, the graduating student can better understand which factors matter most and which offers fair the best with respect to the top selection factors.

Utilizing AHP for assisting in a cross-functional evaluation of WMS vendors is a good step in group decision making. Members of the group must understand that the results are perceptions of the group, and as such, reflect the members of the group. Adding other employees may alter the results. Moreover, the results are a reflection of the criteria and vendors within the analysis but do not consider criteria or suppliers outside of it, such as the following.

- Determine the goal of AHP analysis and the best supplier (not just lowest cost or quickest delivery).

- Identify primary criteria and associated weights.

- Buyers must consider the value of supplier attributes (pairwise) and score each potential supplier on each dimension.

- Combine and generate an overall score.

- Indicate the relative importance of criteria.

- Create a method to prioritize alternatives with multiple criteria.

- Facilitate group decision making.

- Determine if criteria may conflict with one another.

- Attempt to capture subjective and objective criteria.

Figure 14-2 contains the WMS selection criteria, Figure 14-3 contains the WMS vendor comparisons, and Figure 14-4 contains the final WMS vendor rankings that together illustrate the AHP approach to vendor selection. Although the example is simple, more variables and more suppliers could easily be considered. Suppose that the WMS evaluation task team consisted of people from the warehouse operations floor and included a lift operator, inventory clerk, receiving clerk, warehouse section supervisor, transportation clerk, representative from marketing and sales, accounting supervisor, information technology specialist, and a total of three managers from warehouse operations, traffic, and Human Resources. The 11-person team consists of users, frontline employees, managers, and representatives from other departments that may hold some level of stake in the decision.

Four selection criteria and four WMS vendors appear in the example. The first step is to compare one individual criterion to another individual criteria, and repeat the process until all pairwise criteria are compared. Your scale for comparison should start with 1.00 to designate equally preferred. For example, for the criteria labeled Features, when compared to itself it will be of equal importance or equally preferred. Therefore, a 1.00 coefficient is entered into the matrix cell where the row labeled Features intersects with the column labeled Features. You will notice this entry in Figure 14-2. Because this relationship holds true for each criteria compared to itself, the intersecting cells within the diagonal will contain 1.00.

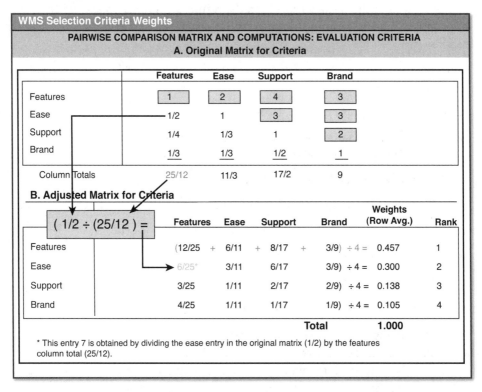

Figure 14-2    WMS selection criteria weights

Because selection of a WMS vendor requires evaluating candidates on the software features, ease of use (ease), support during training, activation and after market (support), and the quality reputation of the vendor (brand), these four criteria were selected for this example. Other criteria could have been selected, and in fact, an AHP analysis comparing individual features could prove useful in evaluating and ranking the most important to least important features.

A measurement scale must be used to assign numerical comparisons between pairs of criteria. In this example, a 10-point scale is utilized with the lowest anchor beginning with 1.00 (indicating "equally important") and the highest score set at 10.00. A perfect 10.00 would indicate that one criteria is "extremely important" over another.

Although the example utilizes a 10-point scale, the group could have utilized, for example, a 1.0 to 100 point scale. In the Original Matrix shown in Figure 14-2, you can see that the WMS Features offered by a vendor is two times (2.00) more important than the software's ease of use or Ease. Below the diagonal you see the reciprocal (1/2) in the intersecting cell for the two criteria. This would indicate that software ease of use is only 1/2 as important as the features offered. After completing the pairwise comparisons for the criteria, the columns are then totaled.

Each cell appearing in a column is then divided by its column total. This equates to the percentage of the cell in terms of the total column value. This is performed for each cell and column total, and the results are entered into an adjusted matrix. Figure 14-2 illustrates that the first cell (1.00) in the Features column divided by the column total (25/12) equals: 12/25. After each adjusted matrix cell is calculated, the final stage for comparing criteria is to average the rows. After conducting the pairwise analysis, the results indicate that the most important criteria in selecting a WMS, in this example, is the features or functionality built into the software (12/25 + 6/11 + 8/17 + 3/9) ÷ 4 = 0.457 wt. As a check, the resulting weights should add up to 1.00.

By now, it should be easy to see that an AHP analysis would best be entered into a spreadsheet so that the many computations can easily be performed. More variables will add more computations.

Having calculated the criteria weights and rankings, the next step is to perform the exact comparison/calculation process again, however, this time comparing vendor-to-vendor with respect to one criterion at a time.

Figure 14-3 illustrates this process and results when comparing vendors based on the features available in their WMS software. Vendor-D ranks first (0.563) and followed second by Vendor-A (0.297).

Figure 14-4 illustrates the Final Matrix analysis and resulting final weights utilized to rank-order the WMS vendors.

**WMS Vendor Comparisons**

### VENDOR PAIRWISE COMPARISON MATRICES AND PRIORITIES

| | Vendor-A | Vendor-B | Vendor-C | Vendor-D |
|---|---|---|---|---|
| **A. With Respect to** Features | | | | |
| Vendor-A | 1 | 5 | 6 | 1/3 |
| Vendor-B | 1/5 | 1 | 2 | 1/6 |
| Vendor-C | 1/6 | 1/2 | 1 | 1/8 |
| Vendor-D | 3 | 6 | 8 | 1 |
| Column Totals | 4.367 | 12.5 | 17.0 | 1.62 |

| | | | | | | Weights | Rank |
|---|---|---|---|---|---|---|---|
| **B. _Adjusted_ Matrix for** Features | | | | | | | |
| Vendor-A | ( .229 + | .400 + | .353 + | .205) ÷ 4 = | | .297 | 2 |
| Vendor-B | .046 | .080 | .118 | .103 | | .087 | 3 |
| Vendor-C | .038 | .040 | .059 | .077 | | .053 | 4 |
| Vendor-D | .687 | .480 | 471 | .615 | | .563 | 1 |

Figure 14-3    WMS vendor comparisons

**FINAL WMS Vendor Rankings**

### COMPUTATION OF WEIGHTS: VENDOR ALTERNATIVES

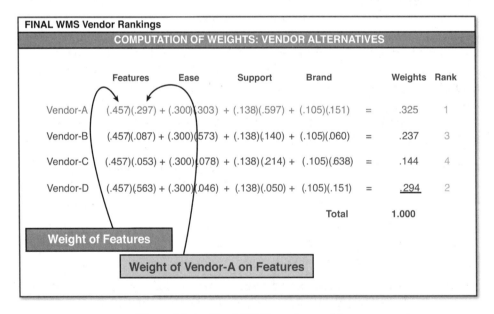

| | Features | Ease | Support | Brand | | Weights | Rank |
|---|---|---|---|---|---|---|---|
| Vendor-A | (.457)(.297) + | (.300)(.303) + | (.138)(.597) + | (.105)(.151) | = | .325 | 1 |
| Vendor-B | (.457)(.087) + | (.300)(.573) + | (.138)(.140) + | (.105)(.060) | = | .237 | 3 |
| Vendor-C | (.457)(.053) + | (.300)(.078) + | (.138)(214) + | (.105)(.638) | = | .144 | 4 |
| Vendor-D | (.457)(.563) + | (.300)(.046) + | (.138)(.050) + | (.105)(.151) | = | .294 | 2 |
| | | | | Total | | 1.000 | |

**Weight of Features**

**Weight of Vendor-A on Features**

Figure 14-4    Final WMS vendor rankings

Final weights are calculated by multiplying the vendor's weight on the respective criteria to the criteria weight resulting from the pairwise criteria comparisons. This is done for each vendor and criteria and summated for the vendor row.

| Features | Ease | Support | Brand |
|----------|------|---------|-------|

**Vendor-A: (.457) (.297) + (.300) (.303) + (.138) (.597) + (.105) (.151) = .325**

Vendor-A took the highest ranking (0.325) followed in second place by Vendor-D (0.294), third by Vendor-B (0.237), and last by Vendor-C (0.144). The results could be placed within a Product Positioning map to illustrate the gaps between vendors on each criteria.

When selecting a WMS vendor there are many considerations that must be made. Some of the most important considerations have been previously discussed. The weighted average-type analysis illustrates that a well-formed evaluation group can provide management with a ranking of selection criteria and a ranking of vendors on the criteria. In the example, Features was a dominant selection criteria followed by ease of use. Both first and second ranking vendors scored high on having the best WMS software features for executing warehouse operations.

## *Hands-Free Voice Technology*

Information technology certainly has helped to reduce errors in the warehouse, and as a healthy consequence, employees and management probably feel better about not committing as many errors. Another technology innovation helping to improve accuracy and speed when handling freight in the warehouse includes hands-free, voice-enabled technology.

Moving from a paper-intensive picking process to a picking by voice technology can bring accuracy to the picking process by reducing errors, some reports indicate, by 80 percent or more. Although variety exists in selecting a vendor, the equipment technology enables the operator to utilize both hands when picking (hands-free), and other than speaking a portion of the barcode for computer verification, the operator is hands-free and eyes-free when picking. Productivity, accuracy, safety, and even employee enthusiasm improves. Voice technology in the warehouse can be applied in most interactions between employees and stock. From cycle counting, replenishment, putaway, picking, and verifying, voice technology can be utilized.

## *Radio Frequency Identification (RFID) Tag and Barcode Technology*

Radio frequency identification (RFID) technology has been utilized for many years and applied in many business settings—from identifying and tracking the smallest SKU measured as pieces to vehicles and 40' intermodal shipping containers. Information is sent via radio waves (frequencies) to an antenna linked to a computer information system. As illustrated in Figure 14-5, a barcode is scanned and the laser light withint the handheld scanning gun records the reflecting laser, information about the product is recognized by the computer information system and revealed to the operator. Each line in the barcode represents specific data associated with the product scanned. There are multidimensional barcodes with the capability to connect operator and computer with large amounts of product information. Barcode labels are easily applied and relatively inexpensive to print, and barcoding is a proven technology that utilizes RF technology.

Research conducted at Michigan State University (2005) discovered the potential gains of utilizing RFID tag technology when compared to traditional barcode utilization through mapping the many processes conducted in three Fortune 100 manufacturer support consumer goods package warehouses. Computer simulation results indicated that by employing RFID tag technology the warehouse operation could experience improved performance time equating to a 58 percent cost per case savings. When human interaction variables were entered into the simulation, there was an additional productivity gain of 4-5 percent. In addition to the findings based on quantitative data, qualitative interviews from all levels of warehouse employees and supervisors within the facilities indicated the most important contributions of RFID tag technology from the human perspective. Figure 14-6 illustrates the findings.

Figure 14-5    RF Technology in the Warehouse

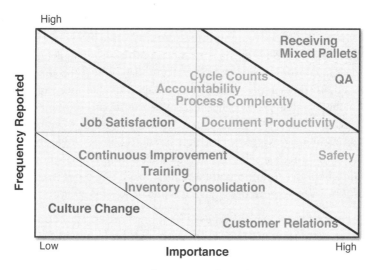

Figure 14-6    Impact of RFID tags for information exchange

Receiving inbound freight appeared to be where the greatest gains in speed and accuracy could be realized by using RFID tags. In particular, mixed pallets of varying SKUs require significantly more time to verify count when they are off-loaded freight. Simulation results and the interview responses concurred. In addition, results of the long interviews indicated that employees believed RFID tags could assist in achieving greater quality assurance, improving cycle counting processes, and simplify other core warehouse processes. Because worker safety and productivity are two of the most important factors in conducting successful warehouse operations, employees believed that RFID tag technology in the warehouse could help to create a safer and a more productive workplace. Factors less likely to be improved through the application of RFID tags included, for example, changes in the business culture, improved customer relations, and the training of employees.

# Integrated Handling/Storage Equipment Technology

In today's most sophisticated distribution centers, you may see zoning indicating high-bay storage operated with an Automated Storage and Retrieval System (AS/RS). Within the same DC, there may be automated guided vehicles (AGVs) traveling to shipping lanes. You may even see piece picking by light systems or by an automated dispensing A-Frame. With all the available technology, the forklift continues to integrate with newer equipment technology to move product throughout the warehouse.

Like the correct selection of a pair of athletic shoes, a kayak, or a surfboard, you must know the intended use and goal of the warehouse handling and storage technology so that the investment returns the anticipated results. The best pair of basketball sneakers would never measure up in a marathon race against a pair of lightweight running shoes specifically designed for distance runners. A kayak designed for paddling on calm and protected waters would have tremendous difficulty attempting to cross a 3-mile bay with a 5-mile head wind. Surfers select their boards based on the conditions of the surf, just like golfers have a club for managing various conditions on the course. Figure 14-7 illustrates that to maintain flexibility to perform many tasks in the warehouse, perhaps a smaller more agile and responsive lift is needed for the job; just like a surfer needs a shorter board to maneuver through and around fast-breaking waves under erratic surf conditions. On the other extreme, to achieve consistent and efficient progress, perhaps a high-bay automated storage and retrieval system or turret lift will be required; similar to the logic that a long-board will provide a smooth and efficient ride when surf conditions are well organized. The point is, equipment must be selected so that it helps to achieve the goals for which it was intended.

The next few sections discuss various types of handling and storage equipment by explaining the characteristics and benefits of each. Although handling and storage equipment can vary in the thousands, you need to understand key applications.

Figure 14-7    Select equipment based on the task. Smaller, more agile equipment may be called for in certain circumstances, whereas larger, more study equipment may be required for other jobs. (Photos courtesy of Marcin Balcerzak/shutterstock, Pensacolasurf.com & Alex McBride, Marcin Balcerzak/shutterstock, and Keith "Kid" Wilkins, Pensacola, FL.)

Similar to selecting a WMS, the decision will likely turn out better by correlating the capabilities or competencies of the equipment to the warehouse goals and to the scope needed within the warehouse. Table 14-1 illustrates this connectivity for specific handling and storage equipment technology. Notice that the type of equipment is listed in the final column. This makes the point that it is important to first evaluate the primary warehouse goals that must be achieved by the equipment, establish the scope and operating conditions within which the equipment will be required to operate, and decide on the capabilities and competencies pertaining to the functional needs of the warehouse operator. Only then should an investment into a specific piece of equipment be considered. Of course, this is a starting point because initial price and training, cost of ownership, payback period, factors pertaining to the vendor and brand (quality and service) and other factors must also be entered into the analysis of equipment technology and vendor selection.

**Table 14-1**     Handling/Storage Equipment Technology

| Warehouse Goals | Scope of Need | Capabilities | Equipment |
|---|---|---|---|
| ■ Reduce putaway time<br>■ Reduce picking time<br>■ Simplify processes<br>■ Improve safety | ■ Routine receipt/ picking<br>■ Full pallet quantities in/out<br>■ Large storage capacity | ■ High-bay capacity<br>■ Narrow-aisle<br>■ Automated<br>■ Integrates with other automation | ■ Automated Storage and Retrieval System (AS/ RS) |
| ■ Improve flow from pick to staging/ loading<br>■ Simplify processes<br>■ Improve safety | ■ Routine movement of unitized/ palletized freight<br>■ Distant travel storage to shipping | ■ Heavy lift capacity<br>■ Multiple cart pulls<br>■ Programed for need<br>■ Integrates with other automation | ■ Automated Guided Vehicle (AGV) |
| ■ Lower operating cost<br>■ Eliminate errors in piece pick<br>■ Speed up piece pick process<br>■ Improve flexibility | ■ High volume order pick<br>■ Piece pick (eaches)<br>■ Varied SKUs | ■ Handles multiple packaging forms<br>■ Dispenses smallest of items with varying product traits<br>■ Available portable and wireless | ■ A-Frame dispensing/ picking system |

| Warehouse Goals | Scope of Need | Capabilities | Equipment |
|---|---|---|---|
| ■ Lower operating cost<br>■ Eliminate errors in piece pick<br>■ Speed up piece pick process<br>■ Ability to integrate technology | ■ High volume SKU piece pick<br>■ Highly varied SKUs<br>■ Integration with other equipment technology | ■ Visual light assistance<br>■ Accommodates many break-pack SKUs<br>■ Accommodates zone picking | ■ Pick-to-light system |
| ■ Lower operating cost<br>■ Speed up cross-dock freight<br>■ Zero errors | ■ Cross-docking to multiple locations<br>■ Varied locations | ■ Visual light assistance<br>■ Handles multiple outbound locations | ■ Put-to-light system |
| ■ Lower operating cost<br>■ Reduce operator travel distance and spend more time picking<br>■ Ability to integrate with other technology | ■ High volume SKU piece pick<br>■ Highly varied SKUs<br>■ Integration with other equipment technology | ■ Moves product to picker<br>■ Indicates bin and quantity to pick utilizing light-pick technology<br>■ Multiple items move to picker simultaneously | ■ Carousel system |
| ■ Lower operating cost<br>■ Reduce operator travel distance and spend more time picking<br>■ Ability to integrate with other technology | ■ High volume SKU piece pick<br>■ Highly varied SKUs<br>■ Integration with other equipment technology | ■ Moves product to picker<br>■ Indicates bin and quantity to pick utilizing light-pick technology<br>■ Multiple items move to picker simultaneously | ■ Robotic racks |

| Warehouse Goals | Scope of Need | Capabilities | Equipment |
|---|---|---|---|
| ■ Improved flow from pick to staging/ loading<br>■ Simplify processes<br>■ Improve safety<br>■ Reduce operator travel distance and spend more time picking<br>■ Ability to integrate with other technology | ■ Piece, case, or pallet pick<br>■ High speed, volume, and velocity product movement<br>■ Multiple industry appeal | ■ Handle high volume<br>■ Integrate with other equipment technology<br>■ Unlimited applications | ■ Conveyor system |

## Automated Storage and Retrieval Systems (AS/RS)

Hyundai's Montgomery, Alabama, automobile plant utilizes an AS/RS to store automobile body parts. For example, after a door is formed by a process that stamps the part from a cut steel plate, the part exits the stamping process via a conveyor belt. The conveyor system technology is integrated with the AS/RS, and together the door is moved from stamping to materials inventory. High-bay storage is utilized to store the many parts at the start of the body assembly process.

AS/RS equipment technology is also utilized in the high-performance Amway Distribution Center in Ada, Michigan. Loaded pallets are received and placed into high-bay storage. The AS/RS selects the optimal location and stores each pallet without operator interaction. Aisles are narrow and maximize the storage capacity availability for large volume palletized product. During picking, the integration of Amway's AS/RS with its automatic guided vehicles (AGVs) enables a steady flow of palletized product from picking to the shipping lane.

As illustrated in the examples and shown in Figure 14-8, AS/RS equipment technology has some important capabilities that may meet the goals and scope of warehouse and distribution centers in various materials and goods industries. Interestingly, Georgia Southern University's library has a multilevel AS/RS for efficiently storing books. Students along with university librarians utilize computer terminals to request a book from storage. The AS/RS locates the book and transports a bin to the "picker." Along with other books catalogued together, the requested book will be located within the bin. Consolidating books into high-bay storage allows the library to offer more variety in services to students in the square footage space gained.

Figure 14-8    Automated Storage and Retrieval System (AS/RS)

AS/RS capabilities or competencies include offering high-bay capacity. Volume is important so that the high fixed cost of the equipment technology can be spread over a large number of units. Narrow aisles within which the AS/RS operates helps to reduce the storage cost associated with square footage set aside for the system and maximum expected inventory. Personnel resources are minimized for the putaway activities assigned the AS/RS. Safety is improved because there is less interaction between employee and machine. Labor costs in the putaway process are significantly reduced compared with a process utilizing more traditional employee operated forklifts.

## Automated Guided Vehicles (AGVs)

AGVs are computer programmed self-guided vehicles that perform product movement tasks for materials and finished goods. Although the Amway DC uses AGVs for moving palletized freight from the AS/RS to shipping, Hyundai utilizes AGVs for transporting the auto chassis combined with transmissions and engines. This occurs at the start of the assembly process as automated machinery unloads trailers and places the heavy components onto an awaiting AGV.

AGV equipment technology is utilized at the end of a chemical production line where bags filled with product pellets are palletized using an automated palletizer. when completed, the palletized product is placed onto an AGV and moved to finished inventory or directly to shipping for export.

Warehouses realize greater levels of workplace safety. AGVs operate with optical scanners that command the AGV to immediately stop if approaching an unexpected obstacle. Fewer forklifts are required in the area, and a steadier flow of material and goods can be achieved from receiving to picking and shipping. As shown in the chemical product example, AGVs enable a steady flow of product from the end of production to the plant-supported warehouse. AGVs may be best employed for heavy and potentially bulky product, and for processes that require product transport from one area of the facility to another.

## Piece Picking

Although AS/RS and AGVs are suitable for large scale and large unit movement, piece pick or picking individual pieces (also called *eaches* in some companies) from opened cartons (break packs) to fill small orders for multiple SKUs requires a different equipment technology strategy and selection. A-frame, pick-to-light, and carousel systems handle such a task—some with total automation and others with automated assistance.

## A-Frame

A-Frames utilize gravity and electronic mechanisms to dispense SKU pieces when ordered. In the form of an "A," products are stacked in vertical channels of the A-Frame. A conveyor situated between the legs of the "A" captures the dispensed product. Some processes utilize cartons or totes riding the conveyor through the legs to collect the dispensed product. Many orders containing few lines of varying SKUs are well suited for A-Frame equipment technology. An A-Frame is utilized in a home shopping DC to pick small cube SKUs that will be shipped parcel post to consumers. Just about anything you could envision fitting into a common vending machine can be picked via an A-Frame system. High-speed dispensing allows for the picking of multiple items at a time. Products may be deposited into a shipping carton that then is conveyed over a scale to verify that the items within the carton total to the weight of the items that appear on the order. The carton may be conveyed directly to shipping or to another area in the warehouse where additional items ordered are added to the carton.

## Pick-to-Light and Put-to-Light Systems

An alternative to the A-Frame equipment technology, a pick-to-light system requires an order filler to move to the location of the SKU split case. Open cases align in horizontal rows, whereby, an order picker will communicate to the picking function within the WMS that a specific order is being picked. After the operator affixes a barcode to a tote or carton for shipping, the scanned barcode initiates the picking of SKUs by piece. A light indicator illuminates at the location of the first item to be picked. A display near the light indicates the number of pieces to pull from the bin and place in the shipping tote. When the item is picked, the operator pushes a button to confirm, and the light turns off.

The light indicator for the next SKU then illuminates, and the picker moves to the item. The process continues until the last light indicator displays a command that the order is complete and the operator can place the shipping tote on the conveyor for shipping.

Many high-velocity retail distribution centers, such as Wal-Mart and CVS Pharmacy, utilize pick-to-light systems for managing pick-pack operations. Some employ light-pick technology to manage cross-dock pick-pack orders. Put-to-light or pack-to-light systems enable operators to break open received cross-dock cartons, and using put-to-light technology, place the correct quantity of pieces for each SKU into a carton or tote scheduled for shipping to a specified store location.

## Carousel Systems and Robot Technology

Carousel systems and sophisticated robotic technology operate under the premise that it is most efficient to move the inventory item to the order filler instead of moving the order filler to the item. Michigan State University's warehouse provides for more than 60,000 students, faculty, and staff. A carousel system consisting of several modules is used to move SKUs, such as a pack of pencils or tape, to the order filler for packing small orders destined for an office on campus. The system is equipped with a light-picking system that indicates which location of the bin and quantity to pick corresponding to the ordered SKU.

For a pick-pack operation of a high-volume Internet retailer, an army of robotic dollies move under specific racks containing the SKUs for picking and then lifts each rack and transports it to the order filler. A laser illuminates the bin location for picking. Robots wait in cue until the operator accesses the bin. Speed and accuracy are achieved using robotic systems.

### Conveyor Systems

Piece picking, case picking, and pallet picking operations may leverage the speed, efficiency, and consistency of conveyors to improve velocity and ultimately inventory turns. Overhead conveyor systems are utilized to move auto body parts from storage to the assembly line but are also utilized to move empty cartons to picking locations for consumer packaged goods. Equipped with rollers, tilt trays, push shoes, and optical laser scanners to read smart labels, conveyors are utilized for high-speed sortation and distribution of overnight documents and parcel post packages via air freight carriers. The high-speed conveyor system depicted in Figure 14-9 integrates well with other storage and handling equipment technology.

Figure 14-9    High-speed conveyor system

## Forklifts

Forklifts have been the workhorses in the warehouse business ever since the beginning of the industrial revolution began in the United States (see Figure 14-10). Similar to today's pallet jack, the earliest lifts were designed with a foot pedal to lift freight off the ground just enough to slowly move it around the warehouse loading dock.

Figure 14-10    Lift utilized in today's warehousing

Modern-day competitive supply chain logistics environments require much more than simply moving freight. Now warehouses must help companies compete by moving freight faster, with zero damage goals, and at low operating cost. And just like selecting a WMS, it is highly important to establish how the selected lift machinery's distinctive capabilities meet the goals of the organization and provide for the necessary scope of operating the lift.

The scope of operation or type of work to perform with the lift and the characteristics or features of the lift must be well matched to achieve the desired outcome. Sometimes, a simple low-lift pallet jack or "walkie" can most efficiently provide for the freight movement needs. Walkies or pallet jacks vary in capabilities. Manual walkies hydraulically lift the pallet, but the operator must utilize his own muscle to pull or push the load. However, moving freight longer distances within the warehouse may best be performed efficiently and expediently utilizing a sit-down electric *rider* lift. Most common, riders are versatile, durable, and flexible in the work that can be performed. They enable operators to sit and some enable standing. Depending on the application, standing may be more comfortable for the operator that must routinely move off and on the equipment. Man-up lifts move the operator and freight vertically to the intended bay location, whereas walkies are more limited to the bottom two bay locations.

Fixed cost to purchase, energy cost to operate, maintenance cost, service, and support, and reliability must also carefully be evaluated prior to equipment selection. Total cost to operate and total cost of ownership must be compared with the benefits of increased warehouse space utilization, improved putaway and pick efficiencies, and other improvements to process and control expected with the new equipment.

Pros and cons of energy sources must be considered. Cleaner technology utilizes electric or propane sources. However, traditional gas and diesel fuel, and the use of natural gas options also exist. Pricing and capabilities differ between alternatives.

Operators must select between a traditional lift truck that requires 12-foot aisles to maneuver compared to a narrow aisle option. Space savings can add up to more storage availability for product. Double- or single-reach capabilities are available. High-capacity, two-deep rack storage may be most efficient for space utilization. Lower capacity levels may require less space, and a single reach truck would work well and cost less. Swivel forks are equipped on turret lifts. The device enables the operator to swivel the pallet without moving the entire lift. This reduction in movement enables narrow aisle operation.

Forklift attachments come in multiple configurations. Again, the type of freight and handling expected will dictate selection of the attachments. As described in Chapter 9, vacuum-suction attachments may be used to affix to rolls of product and move without damaging the exterior of the freight. Roll clamps are also designed for the job. Lift hook attachments are available for latching onto specialized freight. Fork extensions can be attached to move multiple pallets. Attachments exist for lifting and moving multiple drums containing bulk material.

Break bulk material requires specialized forklift attachments called *coil rams* to move nonpalletized steel coils from the warehouse transit shed onto a flatbed trailer. Although low-density dry cereal is easily moved and stacked utilizing a clamping device that grabs onto the slip-sheet and pulls the unitized stretch-wrapped cases onto the lift forks, attachments can also include platforms and cages to lift supplies and others to lift personnel. Although the attachments can seem endless, the determining factor is the material or product characteristics in the end that determines the necessary lift attachments to procure.

## Summary of Key Points

Selecting a WMS, selecting other technology enablers, and selecting the appropriate material handling and storage equipment are critical steps in warehouse management. The WMS helps to optimize and schedule pick waves, manage zone picking strategies through employee assignment and generates shipping documents. Key elements to consider when selecting a WMS include

- Operational goals and corporate goals must be well defined and understood before selecting a WMS.

- Processes and the activities making up the processes must be taken into consideration during vendor evaluation.

- Identify the warehouse or distribution center's needs so that the most appropriate WMS system can be selected to provide the most appropriate functionality.

When selecting handling and storage equipment, first evaluate the primary warehouse goals that must be achieved by the equipment, establish the scope and operating conditions within which the equipment will be required to operate, and decide on the capabilities and competencies pertaining to the functional needs of the warehouse operator. Only then should an investment into a specific piece of equipment be considered.

## Key Terms

- A-Frame

- Active Picking Location

- Advanced Shipment Notifications (ASN)

- Analytical Hierarchy Processing (AHP)

- Automated Guided Vehicle (AGV)
- Automated Palletizer
- Automated Storage and Retrieval System (AS/RS)
- Break Bulk
- Carousel System
- Coil Rams
- Consolidation
- Conveyor System
- Cross-Docking
- Cycle Counting
- Distribution Center (DC)
- Downstream
- Drop Trailer
- Enterprise Resource Planning (ERP) System
- Freight Costs
- Hands Free Voice Technology
- High-Bay Storage
- Hundred-Weight
- Inventory Velocity
- Less Than Truck Load (LTL) Optimization Analysis
- Physical Inventory
- Pick to Light and Put to Light Systems
- Picking
- Product Positioning Map
- Profit Margins
- Putaway
- Radio Frequency Identification (RFID) Technology
- Reserve Picking Area

- SKU

- Slip-sheet

- Stretch/Shrink Wrap

- Supply Chain

- Tariff

- Task Interleaving

- Transportation Management System (TMS)

- Walkie

- Warehouse Management System (WMS)

## Suggested Readings

Akerman, K. B. (1997, 2012), *Practical Handbook of Warehousing*, 4th ed., Chapter 49, Chapman and Hall, New York, NY.

Bowersox, D.J., D. J. Closs, S. B. Keller, and A. D. Ross, (March 1, 2005), "Radio Frequency Identification Application Within the Four Walls of a Consumer Package Goods Warehouse," report for participating companies, Michigan State University.

Harry K.H. Chow, King Lun Choy, W.B. Lee, and K.C. Lau (2006), "Expert Systems with Applications," Vol. 30, No. 4: pp. 561–576.

Nynke Faber, René (Marinus) B.M. de Koster, and Steef L. van de Velde, (2002) "Linking warehouse complexity to warehouse planning and control structure: An exploratory study of the use of warehouse management information systems," *International Journal of Physical Distribution & Logistics Management*, Vol. 32 Issue: 5, pp.381–395.

Petri Helo and Bulcsu Szekely, (2005) "Logistics information systems: An analysis of software solutions for supply chain co-ordination," *Industrial Management & Data Systems*, Vol. 105, No. 1: pp. 5–18.

John F. Kros, R. Glenn Richey Jr., Haozhe Chen, and S. Scott Nadler, (2011), "Technology emergence between mandate and acceptance: an exploratory examination of RFID," Vol. 41, No. 7: 697–716.

# 15

# UNIQUE FUNCTIONING AND UNIQUE MATERIALS WAREHOUSING

## Introduction

Globalization and product variety often drive a need for warehousing capabilities greater than those provided by conventional warehouses. Functional requirements and unique characteristics of goods may dictate the use of specialty warehouses. Product control and security can be facilitated via the use of specialty capabilities provided by bonded warehouses and Foreign Trade Zones. Unique handling requirements and value-added services could also be provided by specialty warehouses. Special handling facilities can also be engaged to assist with reverse logistics requirements.

## Unique Functioning and Unique Materials Warehousing

To label or categorize a warehouse as "specialized" requires some definition and description. Because of the nature of varying products managed and services rendered by warehouses and distribution centers, all storage and distribution operations have elements of specialization. This chapter describes specialty warehousing by the function a warehouse serves and then by the unique nature of the materials and goods handled and stored.

## Functional Specialization

Supply chains, by nature of multiple constituent roles within the many links, nodes and echelons, require inventory to be transported, stored, and serviced (including various forms of product manipulation) to support firms in producing and marketing products.

Strategic and functional contributions of specialized warehouses to achieve their varying critical roles may include facilitating

- Entry of foreign goods into a country
- Financial strategies of client supply chains
- Product flow strategies

Global marketing and trade certainly has elevated international buying and selling activities. Improvements in logistical infrastructures worldwide, advances in transportation equipment, and refinement in supply chain logistics planning and strategizing have made warehousing's role increasingly critical for carrying out the missions of organizations.

## Import/Export Warehouse and Cross-Dock

Regardless of the import country, product entry is dependent on many factors. Import customs agents may require the physical inspection of inbound freight. Although security checks may have taken place during the export process, freight may require inspection because of documentation inconsistencies, the nature of the shipper or export country, the sensitivity of cargo being imported, random selection by customs, or other reasons. Food and drug administration officials within the importing country may require cargo inspection and sampling. In addition to policing and legal requirements to inspect imported freight, much of the cargo entering into a country may not yet have a buyer.

Upon discharging cargo from a merchant ship, it must receive customs clearance prior to being sold in the domestic marketplace. For most imports, it is in the best financial and security interest of shippers and consignees to expedite the clearance process so that freight dwell time is minimized in transit. Foreign entry delays have a rippling effect down the supply chain causing the need for increased inventory to cover sales during the delayed processing time.

When cleared, imported freight may continue through the supply chain to its ultimate destination. For containerized products this may mean a drayage company may pick up the freight for delivery to the rail yard for intermodal transport to the consignee. Other ocean shipping containers may require unstuffing or breaking out the cargo within the container. Contents may belong to one or multiple customers, and a third-party logistics (3PL) provider may be hired to unload and distribute the product on varying carriers.

Marine terminal dockside transit warehouses may be utilized to accomplish customs and other required inspections. They may also provide protection for break-bulk general cargo either palletized, crated, or otherwise unitized. For instance, after custom-cleared, steel coils may then be loaded on trailers that are driven through the warehouse to limit the exposure of the coils to elements of weather.

Beyond marine cargo transit warehouses, an importer may hire the services of a freight forwarder to oversee the removal of the ocean container contents. When removed, freight destined for a specific client may be separated from the other content and loaded onto a trailer for final transit to the destination location.

Both the dockside transit warehouse and the freight forwarder managed warehouse are cross-docking facilities that specialize in managing and physically handling imported freight for redistribution to end customers. Reverse the process and the cross-dock becomes a handling facility supporting the freight handling and legal clearance needs for exporters. Export warehouses may perform stuffing or loading contents from trailers to ocean shipping containers. Special bracing or other dunnage used to secure the cargo while in transit may be installed by the freight forwarder. Ships pitch and roll while in open water, and cargo may shift during transit.

## Bonded Warehouses and Foreign Trade Zones (FTZ)

Bonded warehouses and Foreign Trade Zones (FTZ) are often called upon to provide special warehousing for import freight. Both allow for the warehousing of freight that is under customs rule or scrutiny. Bonded warehouses enable the postponement of import duties until the freight is entered into the marketplace. For an importer who has not yet sold the imported cargo, the advantage is that it can defer duty payment until a buyer is located, and only when the freight is sold and enters the market will duties be assessed and paid. If import cargo is sold to a buyer requiring the shipment to be exported from the custom bonded warehouse, duties may be eliminated on the re-exported freight.

Although bonded warehouse activities may enable the repackaging of freight, break-bulking of freight, and several other restricted activities, a FTZ enables much greater flexibility in managing the freight after entered into the facility. FTZ freight is not considered to be within customs territory, but bonded warehouses are, and a customs entry is completed in this case. For a FTZ, import duties are assessed and paid only on the product entering into the domestic market place. For example, any freight that is destroyed within the FTZ or re-exported is not assessed duty. Whether damaged or waste, bonded warehouse freight requires duty payment based on the tariff rate and freight value when entering the warehouse.

FTZs enable the manipulation and manufacture of product entered. This becomes important for importers who, through the manufacturing process, can change the identity of the product and thereby be obligated to pay the rate of duty assessed on the new product description when entering the market. *Inverted tariff* is the term that pertains to moving from a higher tariff to a lower tariff when an original product's identity is changed within a FTZ. The lower tariff would be associated with the newly manufactured or assembled product. Of course, this assumes that the new product identity is attached to a lower tariff rate by customs rule.

Although there are many similarities and differences between the benefits of utilizing a customs bonded warehouse compared to that of utilizing an FTZ, the restrictions are greater for the bonded warehouse selection. FTZ status is granted for a designated area that may include physical warehouse structures, as well as the land area of the facility (that is, outside and inside storage and manufacture). The FTZ application process may take significant time for approval. Among other requirements, the applicant must

- Justify the need for the FTZ.

- Project the economic impact of the proposed zone.

- Provide evidence that the proposed location does not have an FTZ suitable for the applicants' needs.

FTZ operations may service bulk, break-bulk, and finished goods storage and manufacture. Steel mills, auto assembly plants, freight and package shipping ports (air, water, and land), electronics, and other manufacturing plants, food processing facilities, and a whole host of additional industries find operating within a FTZ to be of strategic importance. A good place to find more information about U.S. FTZs is the U.S. Foreign-Trade Zones Board (www.ia.ita.doc.gov/ftzpage/). The Board provides information on applying for FTZ status, information about existing zone and subzones, and other information to help in the understanding of using or operating a U.S. FTZ. The National Association of Foreign-Trade Zones (NAFTZ.org) is a not-for-profit organization with members whose interest is in U.S. Foreign-Trade Zones. Title 19, Chapter 1A, sections 81a-81u of the U.S. Code contains the FTZ Act, and Title 15 of the U.S. Code of Federal Regulations Part 400 contains the FTZ regulations.

After an approved application is received, the next step is to apply for activation. Activation may also take significant time and require, among other things, a structured security plan in place for protecting the freight from leaving the zone without filing a customs entry or other customs documentation required to move or eliminate FTZ import freight. Customs is assigned to protect the financial interest of the country (that is, the collection of duties and taxes and ensuring quotas are not violated). Import freight entered into the FTZ that comes up missing and without legal documentation indicating its legitimate disposition is suspect. Customs may levy heavy penalties against the FTZ operator's bond for each infraction and until a resolution is rendered.

Activation requires background checks for personnel working within the FTZ. Signage must be posted to communicate that the area is designated as a foreign trade zone. An operating manual must be drafted and approved by customs. In it the manual must contain the processes and procedures to be performed within the zone. It must include complete descriptions of the procedures for security, receiving, storing, and other work to be performed; inventory tracking and assurance; and shipping. It includes example documents required and governing the activity as products enter, remain within, and exit the zone.

A public warehouse located in Memphis, TN, decided to expand its corporate resume and service offerings by applying for FTZ status and operation. FTZ regulations enable the activation of sections of a total designated zone. Moreover, regulations enable foreign freight and domestic freight to co-mingle within the same area of the FTZ warehouse site. This enables significant operating flexibility for a public warehouse. As the FTZ business grows, the appropriate space allocated for the import product can expand within the activated portions of the zone. However, the empty space within the FTZ can be utilized for any product or warehouse client. Bonded warehouses, however, are restricted to storing only foreign imported goods and are not allowed to co-mingle them with domestic freight.

Decisions of when to use a bonded warehouse versus an FTZ should be based on the expected activity to be accomplished within either. An FTZ strategy worked well for an importer of raw fruit. When entered into the FTZ, the raw fruit was processed and canned. The inverted tariff was reduced to zero; because although the raw fruit had a 100 percent import tariff, the canned fruit had no import tariff when it cleared through customs for sale in the U.S. market. This would not have been achieved in a customs bonded warehouse.

## Unique Materials and Goods

Some materials and goods require specialized handling and warehousing. Live product, such as lobsters harvested off the north Atlantic coast, is warehoused in saltwater tanks and distributed live overnight to restaurants. Clearly, not all distribution centers or warehouses are alike because not all products are alike and require unique handling during warehousing and distribution.

### *Refrigerated and Temperature Controlled*

Poultry is transported via motor carrier from processors located in the Southeastern United States to the Port of Pensacola, Florida, where it will be warehoused and ultimately exported on a refrigerated cargo vessel. Pensacola's port is equipped with a cold storage warehouse that does more than just keep frozen chicken frozen. Fresh unfrozen chicken can be flash frozen right inside the port's warehouse.

Chocolate candy bars also require temperature controlled storage. However, the refrigerated warehouse must regulate temperatures to keep the candy from freezing and to protect it from high temperatures and melting.

Fresh flower supply chains rely on the speed of airfreight services to assure the perishable flowers arrive to customers worldwide. Temperature controlled warehouse locations package fresh flowers in flat boxes and in water-filled containers for those most sensitive.

## Bulk and Break-Bulk Materials Warehousing

Many ports are equipped with silos to store cement off-loaded from bulk ocean-going vessels. Upon order receipt, the cement is loaded into rail hopper cars or motor carriers equipped with hopper trailers and transported to customers. Other dry bulk materials commonly warehoused in silos include grain, woodchips, coal, and other agriculture and building raw materials. Liquid bulk oil is transported 800 miles via the Trans-Alaskan pipeline from Prudhoe Bay to Valdez, Alaska, where it is stored in holding tanks until loading on crude oil bulk tankers heading for refinery processing.

Break-bulk or general cargo may be palletized or nonpalletized, banded together, and contained in shipping drums or bags. Many times the type and weight of the cargo limits the stacking ability. For example, coiled carbon steel bands used to produce auto body parts and crated pulp-wood products used as material for the production of diapers to food filler require protection from weather elements. Large bags of grain and bags of plastic-material pellets may also be considered break-bulk cargo and require covered warehouse storage.

## Other Specialized Warehousing

Specialized equipment and expertise is required to operate a packaging services warehouse. Contract services may include creating customized mixed pallets from multiple single-item full pallets. Value-added services may also include building store-ready displays complete with product assortment and ready for displaying on the retail floor directly upon delivery.

GENCO manages returns processing for major retailers and manufacturers. Reverse logistics, or the movement of product back up the supply chain, can be difficult to manage due to the loss of economies of scale, mixed product returns, and overall lack of standardization and common forward supply chain management conditions. In the special handling facility, products are returned for redistribution, repair, refurbishment, or recycling. GENCO's service relieves the retailer or wholesaler from the responsibility of handling returns. Actually, GENCO has a service offering to manage product recalls for clients. Anyone in the warehousing and distribution business knows the importance of quickly responding to product recalls. Sometimes, the speed and accuracy of managing recalls and returns can mean the difference between the life and death of a brand.

Fishing and leisure boats are moved via a heavy forklift from water and placed in a high-bay storage warehouse. As seen in Figure 15-1, the facility offers inside warehousing in racks or outside warehousing on two-tiered racks or individual yacht cradles or trailers. Heavy lift machinery is required to manage the receiving and putaway of the boat.

Figure 15-1    Fishing leisure boat high bay storage

An unique floating warehouse and manufacturing facility is located inside the Global 1200 ocean-going vessel. The massive deepwater derrick pipeline vessel and crew of more than 200 when operational fully assembles the pipeline within the ship and extrudes the finished pipe to some of the deepest parts of the ocean floor. Forklifts, materials, and other equipment and tools required to do the job are warehoused on the ship. Pipe inventory is transloaded from barge and cross-docked directly into production.

Specialized warehousing exists everywhere. Goals, scope of need, capabilities, and cost of operation and ownership (total cost of ownership) must all be well defined and weighed into the selection of a specialized warehouse and provider. Sometimes, the functioning of the warehouse within the supply chain is of greater influence than the type of product itself. But, as can be seen in the Figure 15-2, sometimes product characteristics demand a specific type of warehouse or distribution center option.

Figure 15-2    Global 1200 Deepwater Derrick pipeline vessel (complete on-board warehouse and production)

## Going Global

With the proliferation of global marketing exchanges, the need becomes greater for warehouse and distribution centers that can respond to demand changes anywhere around the globe. High-tech warehousing of today has made supply chains and corporate board rooms more dependent on warehousing for adding value and quality while helping to reduce the overall total supply chain cost.

Strategic supply chain managers must understand that differences exist in languages, cultures, and transportation and distribution infrastructures. For the most progressive economies, the differences may be minimized. However, to reach the labor forces and markets (consumer or industrial; product or service) in some developing regions, a greater investment in the critical components of the supply chain infrastructure will be required.

Densely urban area populations require attention to roadways and distribution points for providing products to meet rising demand. Struggling questions will likely include, but go well beyond the following:

- How and where should warehousing and distribution points be located?

- How can roadways and populations support the required speed and capacity of distribution to meet customer demand expectations?

- Can existing warehouse strategies be successful, or will completely new strategies be necessary?

- How can warehouse and distribution operations adapt to supply for the needs of varying populations living under vastly differing economic conditions?

## Summary of Key Points

In some respect, every warehouse has an element of specialization about it to accommodate the specific product's nature and requirements. However, there are instances in which true specialized warehouses are needed to accommodate functional needs or unique product characteristics. Bonded Warehouses and Foreign Trade Zones are two examples of entities that provide product control and security in the supply chain when needed. More straightforward examples of specialty warehouses are those that accommodate products needing temperature-controlled environments and those that facilitate reverse logistics.

## Key Terms

- Bonded Warehouse

- Break Bulk

- Cold Storage Warehouse

- Cross-Docking

- Distribution Center (DC) Dockside Transit Warehouse

- Drayage Company

- Dunnage

- Dwell Time

- Economies of Scale

- Export Warehouse

- Freight Forwarder

- Import Freight

- Import/Export Warehouse Inbound Freight

- Intermodal

- Inverted Tariff

- Marine Cargo Transit Warehouse

- Motor Carrier

- National Association of Foreign-Trade Zones

- Public Warehouse

- Reverse Logistics

- Specialized Warehouse

- Supply Chain

- Third-Party Logistics (3PL) Provider

- Total Cost of Ownership

- U.S. Foreign-Trade Zones Board

- Value Added Services (VAS)


## Suggested Readings

Akerman, K. B. (1997, 2012), *Practical Handbook of Warehousing*, 4th ed., Chapter 40, Chapman and Hall, New York, NY.

National Association of Foreign-Trade Zones (NAFTZ.org).

Swenson, D. L. (2000), "Firm nd]outsourcing decisions: evidence from U.S. foreign trade zones," *Economic Inquiry*, Vol. 38, No. 2: 175-189.

Title 19, Chapter 1A, sections 81a-81u of the U.S. Code.

Title 15 of the U.S. Code of Federal Regulations Part 400.

U.S. Foreign-Trade Zones Board (www.ia.ita.doc.gov/ftzpage/).

# GLOSSARY OF KEY TERMS AND DEFINITIONS

**A-Frame Systems**: Automated product dispensing mechanisms to assist with product picking. A-Frames utilize gravity and electronic mechanisms to dispense SKU pieces when ordered. In the form of an "A," products are stacked in vertical channels of the A-Frame. A conveyor situated between the legs of the "A" captures the dispensed product. Some processes utilize cartons or totes riding the conveyor through the legs to collect the dispensed product.

**ABC Analysis**: A categorization method that prioritizes inventory into categories with A being the most valuable/critical/time-sensitive and C being the least.

**Accessorial Charge**: Carrier freight charges that pertain to a special service to assist customers during the transport and delivery process. Temporary storing of freight or delivering to a residential location may warrant a special charge.

**Accounts Receivables (A/R)**: Accounts receivables are money owed by entities on the sale of products or services. A/Rs are legally enforceable claims for payment to a company by its customer/clients for goods supplied. An A/R is an accounting transaction dealing with the billing of a customer for goods and services that the customer has ordered.

**Accumulation**: A buildup of inventory levels.

**Active Product/Picking Area/Location**: Active picking areas contain product that is immediately available for filling orders.

**Activity or Process Performance Index**: A Personnel Utilization Metric defined as (Units employee serviced) (Standard time per unit) / (work time for the employee).

**Advanced Shipment Notification (ASN)**: A notification sent to a customer containing information about a pending delivery. The ASN serves to provide shipping information, such as quantity and expected delivery date, and is usually facilitated via electronic data interface (EDI).

**Aggregate Forecast**: Predicting the demand of a series of stock keeping units at, for example, a category level as opposed to an item level forecast.

**Analytical Hierarchy Processing (AHP)**: A tool that can assist in selecting a Warehouse Management System (WMS). AHP utilizes pairwise comparisons between selection criteria and then between vendor candidates/WMS software on each of the selection criteria.

**Allocation**: A method that assigns available resources in a prioritized manner.

**Anticipatory Inventory**: Inventory ordered and held in stock to fulfill future expected demand.

**Area Picking**: A picking strategy where the picker is assigned to an area and picks only product from that area of the warehouse.

**Area Zoning Regulations**: Regulations that dictate occupancy requirements and restrictions within a designated zone (city, county, state, and so on).

**Assortment**: Refers to the separation of inventory items in accordance to storage or delivery priorities or constraints.

**Automated Guided Vehicles (AGV)**: Computer programmed self-guided vehicles that perform product movement tasks for materials and finished goods.

**Automated Palletizer**: Material handling equipment that automates the process of building pallets in a warehouse. This capability not only can speed up the process, but can also increase workplace safety aspects for warehouse employees.

**Automated Storage and Retrieval System (AS/RS)**: Consist of a variety of computer-controlled systems for automatically placing and retrieving loads from storage locations. AS/RSs are typically used when a high volume of goods is moved into and out of storage, space constraints exist; the respective process adds no value added to the product; and accuracy is critical.

**Available Warehouse Space**: The space in a warehouse that is open for product storage. Some space within a warehouse is occupied with administration offices, sanitation, and other functions that do not entail product storage space and limit the available warehouse space for storage.

**Back Haul**: The return travel leg of a transport carrier from its destination to its originating point. This can be loaded or empty.

**Back Orders**: This occurs when an item or product is not available when requested. The requirement for the order remains in the supply channel to record its unfilled need. Back orders can signal operational process inefficiencies or the use of suboptimum forecasting.

**Batch Picking**: Pulling items from stock to fill multiple orders simultaneously. The orders are batched together to reduce the number of trips required to the same location.

**Bid**: A submission made in response to a request for proposal (RFP), also called a proposal. In the bid, the submitting company explains how it can provide the services requested and at what associated cost. The bid is then screened against stated evaluation criteria by the issuer of the RFP.

**Bill of Lading**: The legal document governing the shipment transport from origin to destination. The BOL contains important information pertaining to the shipper (consignor), receiver (consignee), and carrier. The freight description including classification, weight, cube, billing information, special instructions, declared value, COD charges, signatures, and contact numbers are contained on the BOL.

**Bonded Warehouse**: A type of specialized warehouse. Bonded warehouses are often called upon to provide special warehousing for import freight. They enable the warehousing of freight that is under customs rule/scrutiny. Bonded warehouses enable the postponement of import duties until the freight is entered into the marketplace. Whether damaged or waste, bonded warehouse freight requires duty payment based on the tariff rate and freight value when entering the warehouse. Bonded warehouses are restricted to storing only foreign imported goods and are not allowed to comingle them with domestic freight.

**Bone Yard**: A storage area that contains old, unused items or items in disrepair.

**Bottleneck**: An area or section in a process in which capacity is overwhelmed by demand thereby slowing down overall progress.

**Bracing**: Material used to pad, protect, and secure goods and cargo.

**Break-Bulk**: The act of separating multiple units of materials or finished goods to carry on further transportation and distribution of the units.

**Break-Pack**: Open cartons used to fill orders requiring less than a full case quantity.

**Bulk Picking**: A picking strategy whereby all like items are picked at once for all orders requiring that item. Groupings can also be by endpoint destination. This strategy reduces over all picking time.

**Bulk Pick Line**: Established to support the strategy of bulk picking.

**Capacity Constraints**: Any phenomena that restricts process capacity. Capacity constraints can become process bottlenecks if not properly managed.

**Capacity Utilization**: A quantifiable metric, often identified as a Key Performance Indicator (KPI) that measures and helps manage the level of process output attained versus the maximum process output possible.

**Carousel Picking System**: Carrousel Picking Systems operate by bring the product to the picker.

**Case Fill Rate**: Ratio of the number of cases shipped to that of the total number of cases ordered (case fill rate = cases shipped / total cases ordered). Used as a measure of customer service.

**Cash to Cash Cycle**: Time elapsed between cash spent for a material or service until payment/revenue is received from the sale to the next-in-line customer.

**Coil Rams**: Specialized forklift attachments to help move nonpalletized steel coils from the warehouse transit shed onto a flatbed trailer.

**Cold Storage Warehouse**: These are specialized warehouses providing environmentally controlled space for storing goods requiring low temperature conditions.

**Commodity Rate**: Rate negotiated for regular and guaranteed high volume transport of a single commodity between two points.

**Consolidation**: Combining multiple orders from a common origin and consigned to a common destination to gain efficiencies in transportation equipment utilization, thereby, reducing transportation cost per unit or order and reducing lead-time.

**Containment Storage Rooms**: Specialized storage areas used to prevent product exposure due to safety or security requirements.

**Contract**: A legally binding arrangement or document between two or more parties entered into voluntarily for the purpose of providing goods or services in return for compensation.

**Contract Warehouse**: A contract warehouse operation guarantees the labor, equipment, and service level stipulated within the contract. The resources will be dedicated more highly to managing the client's business than that provided by a public warehouse operator. The contractor would require a guaranteed level of throughput to make the investment worth the contract.

**Conveyor System**: An automated system that brings the product to the picker.

**Cost of Goods Sold**: These are the costs that can be directly attributed to the production or manufacturer of goods that are sold by a company. Costs include materials used and the direct labor applied. Indirect costs are not included.

**Cost Point of Indifference**: The point in a cost or break-even analysis where pursuing either path under consideration will yield the identical cost.

**Council of Supply Chain Management Professionals (CSCMP)**: The leading supply chain management and logistics management professional association for managers and academics.

**Counseling**: Counseling is provided to employees that require more critical direction to alter behavior patterns and as a final step to avoid dismissing an employee.

**Cross-Dock (Cross-Docking)**: Inbound inventory and orders are received, sorted, reconsolidated based on common destination ZIP code zones or customers and shipped without ever being entered into warehouse storage. In this process freight is moved across the dock from an inbound trailer or container to an outbound.

**Customer Satisfaction**: This is a metric used to define the customer's satisfaction with goods provided by a company. It is expressed as a percentage and is often designated as a key performance indicator (KPI).

**Cycle Counting**: The physical counting of designated SKU inventory on an ongoing basis as opposed to conducting only physical inventory annually.

**Cycle Stock**: Inventory that is expected to be sold during a given order period (cycle).

**Customs Duties**: A customs duty is a tariff or tax on imported goods.

**Deconsolidate/Deconsolidation**: To break-bulk material.

**Demand Forecast**: Projecting the future demand of a product to plan for its accommodation and to ensure its availability when required. Automated systems exist to assist in demand forecasting.

**Demand Patterns**: Identifiable patterns in the demand of an item or service. Examples of demand patterns are those that are seasonal or keyed to the occurrence of a certain event.

**Demurrage/Detention**: Charges levied by the carrier or equipment leasing company for a shipper retaining the transportation equipment past the agreed upon time stated in the contract.

**Designed Warehouse Space**: The total amount of space originally built into the warehouse to house all activities' storage related or unrelated to inventory storage.

**Distribution Center (DC)**: A location that stocks goods/products for temporary storage prior to being transported to the customer. A distribution center is a basic part of a supply chain and is often called a warehouse.

**Distribution Channel**: The end-to-end path through which a product travels to get from the manufacturer to the end user.

**Distribution Costs**: Distribution costs are the sum of all costs incurred in moving goods from the point of production/manufacturer to the end user.

**Distressed Product**: Product that cannot be sold at its original price. This includes inventory that has reached or surpassed its expiration date, items that are defective, and items that are out of date or out of style. Distressed products are candidates to be considered for the reverse logistics process.

**Dock Bumpers**: Attachments that help to provide protection to the dock from damage due to contact with trucks and trailers. Dock bumpers may also assist in guiding trucks into position.

**Dock Levelers**: A dock leveler is a platform put in place to level the interface point between a dock and a servicing truck. These can be manually emplaced separate material pieces or can be mechanically or hydraulically emplaced integral components of the dock.

**Dockside Transit Warehouse**: A warehouse stationed near a dock to serve as a transit point for cargo off loaded from a water borne carrier.

**Downstream**: Looking toward the next-in-line customers or partners in the supply chain. The ultimate consumer is the furthest downstream supply chain member.

**Drayage Company**: A company that provides transport of containerized cargo, often to ocean terminals and railheads.

**Drop Trailer Process**: Drivers drop their trailers in a designated area or door and hook to another trailer, empty or loaded, and depart the facility. Drop trailer processes may occur during regular shift hours or after hours. Drivers have limited interaction with receiving personnel under a drop trailer process.

**Dunnage**: Material used to wrap, pad, protect, and secure goods and cargo.

**Dwell Time**: Refers to the time goods remain in an in-transit storage area while waiting to be shipped; it is also used to refer to driver wait time during which no activity or productivity is occurring.

**Economic Order Quantity (EOQ)**: The order quantity to place with a supplier that is associated with the lowest total cost when considering only the cost to order and the cost to carry the inventory. EOQ is cycle stock or the amount to order each cycle that will satisfy known demand at the lowest total combined order cost plus inventory carrying cost.

**Economies of Handling**: Achieving process efficiencies by using handling techniques to expedite and minimize product handling requirements.

**Economies of Manufacturing**: Achieving process efficiencies by using manufacturing techniques to optimize the manufacturing process.

**Economies of Production**: Achieving manufacturing cost efficiencies by utilizing long production runs of an item to minimize equipment changeover time between items produced on the same production line.

**Economies of Purchasing**: Achieving cost efficiencies by purchasing in quantities large enough to receive price-break discounts from materials and product suppliers.

**Economies of Scale**: Achieving cost efficiencies by increasing output (producing or providing more product) as fixed costs are then spread over more items.

**Economies of Transportation**: Achieving cost efficiencies by transporting in quantities large enough to receive price-break discounts from transportation carriers. Economies are achieved as carriers can maximize the weight and cube capacity utilization of the equipment.

**Electronic Data Interface (EDI)**: Generically used to describe the electronic interface between two or more business entities. The electronic interface is facilitated via computer systems with compatible software or systems. EDI enables e-commerce and can be used to place orders with a warehouse and to track order status through delivery.

**Electronic Picking Tunnel**: Warehouse automation that optimizes the traditional tunnel picking process. The automated picking from locations is conducted within a "tunnel" that traverses through the rack system.

**Employee Productivity**: In its basic sense, this is a measure of the amount of goods or services that an employee produces or provides in a given period of time.

**Employee Turnover**: Employee turnover and turnover rate is a measure of the rate that a company gains and loses employees. It is a measure of employee tenure. Employee turnover can be used for self-evaluation and for comparison to competitors in the same industry sector.

**Enterprise Resource Planning (ERP) System**: An amalgamation of software capabilities that provides support to and assists in the management of company business processes. ERP systems can identify needed resources, track the status of existing resources, and assist in inventory management.

**Equipment Utilization**: A Cost and Utilization Measure defined as (actual machine operating time utilized / total available operating time).

**Evaluation Criteria**: Criteria designated for use in selecting a contract winner. The evaluation criteria, the priority hierarchy, and respective criteria weighting are transmitted to potential bidders in the Request for Proposal (RFP) document. Examples of evaluation criteria include price and past performance.

**Export Warehouse**: A specialized warehouse used to facilitate the export of goods and products. Compliance, licensing, insurance, and documentation support are common value-added services that can be provided by an export warehouse.

**Fill Rate**: A metric, expressed in terms of a percentage equaling the number shipped divided by the total number ordered.

**Financial Incentives**: States and municipalities may offer financial incentives to entice a company to select their area to locate or build their warehouse. Examples of financial incentives include tax incentives and interest-free or reduced loans.

**First-In First-Out (FIFO)**: FIFO is an inventory management process whereby the oldest items on-hand are issued at the first available opportunity.

**Fixed Costs**: Fixed costs do not fluctuate with increases or decreases in output or demand. Examples of fixed costs include depreciation, insurance, and rent.

**Fixed Order Quantity**: A re-order point based on a standardized (fixed) amount of stock to order each time. The time of order may change as demand changes, but the amount ordered each cycle remains the same or fixed.

**Floor Loading/Loads**: Loading cartons and products directly on the trailer floor to achieve the maximum trailer capacity utilization possible.

**Forecasting**: Projecting the future demand of a product to plan for its accommodation and to ensure its availability when required. Automated systems exist to assist in demand forecasting.

**Foreign Trade Zone (FTZ)**: An FTZ enables flexibility in managing the freight after it enters into the facility. FTZ freight is not considered to be within customs territory. For a FTZ, import duties are assessed and paid only on the product entering into the domestic market place. For example, any freight destroyed within the FTZ or re-exported is not assessed duty. FTZs enable product modification and manufacturing within its confines. Regulations enable foreign and domestic freight to be co-mingled within the same area of an FTZ warehouse. This provides operating flexibility for a public warehouse. If FTZ business expands, the space allocated for the import product can expand within the activated portions of the zone. However, the empty space within an FTZ can be utilized for any product or client.

**Forklift**: A material handling system that comes in various configurations and sizes and can be manually operated, automated, or robotic.

**Forward Picking Area**: A component of a picking strategy. The forward picking area of a warehouse or distribution center is established to locate product/SKUs that are fast movers or frequently requested. A forward picking area is established to reduce the labor required to pick the respective items. A forward picking area is replenished from the reserve area or the bulk area, as appropriate.

**Forward Stocking**: Forward stocking locations can be established to enable the quick issuance of product. Value added services (VAS) such as kitting and packaging are often associated with forward stocking locations.

**Fourth-Party Logistics Provider (4PL)**: A supply chain logistics partner that coordinates the services of multiple third-party providers for a specific customer. The 4PL provides a customer with organization and control of logistics services, information, and costs beyond that provided by an individual 3PL.

**Free on Board (FOB)**: Domestic transportation terms of sale that govern the ownership and liability of goods in transit and assign responsibility for carrier freight payment.

**Freight-All-Kinds (FAK)**: A term used to classify different freight types but that have general transport similarities. Although they may be classified in a slightly higher or lower freight class, the shipping volume per item does not warrant separate classifications. The group of products may be placed in a FAK, for example, class 55.

**Freight Bidding Process**: The process including the RFI, RFP, and RFQ stages of negotiations between carriers and shippers. Critical information is exchanged pertaining to the product to be shipped and service expectations to factors pertaining to the carrier's authorization, ability, and interest to provide transport services. The end goal is to conclude with a successful transportation service agreement between select carriers and a shipper.

**Freight Bill**: The document containing freight charges and indicating payment is to be remitted to the carrier for services rendered.

**Freight Bill Auditing**: The process of comparing the freight charges that are billed to a shipper to the agreed upon rates plus any additional charges stemming from surcharges, detention of equipment and/or driver, or any additional accessorial charge. Auditing may entail reconciling the freight bill with information on the delivery receipt, original order, and bill of lading.

**Freight Consolidation**: Combining two or more orders from a common origin that are destined to the same ZIP code zone and/or consignee to gain economies in transportation and reduced freight charges per unit shipped.

**Freight Costs**: Freight costs are the sum of costs incurred during the process of moving products/goods. This sum can include packaging costs, palletizing costs, loading costs, unloading costs, insurance costs, transportation costs, security costs, and tracking costs.

**Freight Forwarder**: A freight forwarder is an agent or company that manages shipments for companies to move products/goods from the manufacturer to the customer. Most freight forwarders do not own the transportation assets; rather they contract for the appropriate modes of transportation. Freight forwarders work in both the domestic and international markets.

**Freight Payment**: Making payment for carrier services rendered.

**Freight Rate**: The price per one hundred pounds (hundred-weight), per unit, per mile, or per load.

**Full Line Stocking**: This technique minimizes the number of warehouses or distribution centers that a customer has to deal with by stocking and providing the entire line of items required.

**Full Truck Load (TL)**: Typically, one large shipment consisting or a single order or multiple orders transported from a single origin directly to the destination customer. At its most efficient, the orders fill up the maximum capacity available, thereby, weighting-out and cubing-out the entire trailer.

**General Services Administration (GSA)**: A U.S. government entity that manages government assets and establishes government-wide policy.

**Grazing**: A term used to describe employee pilfering of foodstuffs from warehouse stocks.

**Hands-Free Voice Technology**: This technology enables the operator to use both hands when picking. This technology can increase the speed and accuracy of the picking process. Voice technology can be applied in warehouses for cycle counting, replenishment, putaway, picking, and verifying.

**High-Bay Storage**: This storage method is for inventory on-hand that is held in high vertical rack pallet locations. In the most efficient operations, the product may be received in full pallet quantities and shipped in full pallet quantities with no additional handling requirements.

**High Cube**: The stacking or racking of product in large cubic storage and high-bay areas or facilities.

**Honeycombing/Honeycomb**: This is the situation that occurs when stock is removed from warehouse locations and the resulting vacancy creates an unusable space. This reduces the efficiency of the warehouse operation.

**Hundred-Weight**: An industry unit of measure. The term is used to describe materials that have a dollar cost associated with a 100-pound unit of measure.

**Iceberg Principle**: The concept that the obvious problems are only the tip of lurking issues that must be addressed.

**Import/Export Warehouse**: Warehouses established to manage the import and export of goods from a country outside of the warehouse location.

**Import Freight**: Freight that is brought into a different country than its country of origin. This freight is subject to the customs regulations of the receiving country and may be subject to additional control and security measures depending on the nature of the freight (for example, quarantine periods, additional taxes, or duties).

**Importer**: A business entity focused on importing or assisting in the import of goods from an outside country.

**Inbound Carrier**: A carrier transporting product into a distribution center or warehouse.

**Inbound Freight**: Freight transported into and received by a warehouse or distribution center.

**Inbound Transportation Cost**: All costs associated with the inbound transportation of goods to a warehouse.

**INCOTERMS**: (2010) International Commercial Terms published by the International Chamber of Commerce to facilitate terms of trade in a common set of 11 terms for the benefit of exporters, importers, carriers, and others who manage international freight sales and distribution.

**Independent Cost Estimate**: An internal estimate of costs expected for services requested of an external agency; conducted by the requesting business entity to use for comparison purposes against estimates received. Used during the request for proposal or request for the quote process to guard against inflated estimates or low-ball submissions.

**Infestation**: A visual cue in a warehouse that product has been environmentally breached.

**In-Stock Inventory**: Goods that are on-hand in the warehouse and available to issue against customer demand.

**Intermodal**: Transporting freight using more than one mode of transportation such as motor carrier, air transport, rail car, water carrier, and pipeline.

**Intermodal Container**: A shipping container designed to transport freight utilizing more than one mode of transportation. Efficiencies are gained as standardized containers enable the protection and transportation of various freight types while the containers are easily transferable between motor, ocean, and rail carriers. Containers may vary in functionality; however, common intermodal containers include 20', 40', 48', and 53' lengths that may be transported on chasses behind standard motor carrier tractors.

**International Standards Organization (ISO)**: The international organization that establishes, regulates, and verifies international management standards.

**In-Transit Inventory/Stock**: Ownership of inventory passes hands as materials and product move through the supply chain. Consequently, at least one member of the supply chain owns the product while it is in motion from suppliers to customers. In-transit inventory, therefore, is as important to manage as in-stock inventory.

**Inventory Accuracy**: Inventory accuracy is a metric measuring discrepancy between the physical inventories on-hand and the inventory levels reflected in the warehouse records.

**Inventory Carrying Costs**: The costs associated with holding product in stock, including storage and handling costs, taxes, interest, and insurance based on the value of the product and product obsolescence costs.

**Inventory Control Clerks**: Perform scheduled physical inventories and reconcile discrepancies between the actual product in the warehouse and what the system indicates is on-hand.

**Inventory Flow**: The term used to describe the end-to-end path of inventory as it moves from the manufacturer to the end user.

**Inventory Integrity**: An aspect of inventory accuracy that assesses the reliability of inventory records.

**Inventory Management System (IMS)**: An automated system to assist in managing inventory. The IMS can be standalone or part of the EDI suite.

**Inventory Objective**: The planned quantity of a product necessary to be on-hand and on-order to meet customer demand.

**Inventory Reorder Point**: The product stockage point at which replenishment inventory is ordered to ensure availability upon demand. Demand usage and demand forecasting calculate the correct inventory reorder point.

**Inventory Stockout/Stockouts**: The situation that occurs when an item is not available when requested. This is an indication of a process problem.

**Inventory Taxation**: A levy placed on goods stored in a warehouse. Commonly experienced in an import/export scenario.

**Inventory Throughput**: A measure of revenue received as a result of inventory output.

**Inventory Turnover (Inventory Turns)**: Times that the total inventory has been sold throughout the year. Typically, calculated as the ratio of the cost of goods sold to the average inventory; with average inventory calculated as starting inventory plus ending inventory multiplied by ½.

**Inventory Velocity**: This is a measure of the time an item spends in storage from the time of receipt until the time of delivery. A high inventory velocity is desirable. This measure can be attained by dividing the costs of goods sold by the average inventory for the selected time period.

**Inverted Tariff**: This term pertains to moving from a higher tariff to a lower tariff when an original product's identity is changed within a FTZ. The lower tariff will be associated with the newly manufactured or assembled product.

**Invoice**: A document establishing the payment required for services or goods provided.

**Job Shadowing**: The practice of a new or untrained employee following a seasoned employee during the conduct of business as a means of training, cross training, and exposure.

**Just-in-Time Replenishment**: A time-based strategy requiring carriers to deliver extremely time-sensitive shipments to the consignee within a short transit time. Shipments should arrive and be ready for use just prior to the customer stocking-out of a product. Time-based strategies help to reduce inventory levels at the customer and demand highly dependable carriage.

**Kanban System**: Comes from the concept of lean manufacturing and serves to control material flow and inventories. This system improves efficiency and helps reduce the costs associated with the respective process.

**Key Performance Indicators (KPI)**: Quantifiable metrics established to measure and monitor process performance aspects. KPIs can include items such as inventory velocity, order ship time, and inventory turnover.

**Kitting: Product Configuration**: This can be done as a value added service (VAS) in a warehouse.

**Knockdown Cartons**: Unassembled corrugated shipping boxes used to pack/repack items. Product rework areas utilize KDs as they are known to salvage good product, repackage the product in a new carton, and place the carton back into available inventory.

**Labeling**: Product identification information affixed to the item.

**Lead-Time**: The time it takes from the time of order to the time of delivery. Also known as order cycle time.

**Leakers**: Stored product that provides a visual cue of a problem when it begins to leak.

**Lean Six-Sigma**: Lean Six-Sigma is a management practice that uses the principles of "lean" to eliminate unnecessary and no-value-added processes and the six-sigma process improvement approach to improve efficiency and reduce waste.

**Less-Than-truckload (LTL)**: Shipment consisting of an order that fails to utilize the entire maximum capacity available in a truckload size trailer.

**Line Haul**: This term is used to refer to the transport of freight over long distances, usually by truck.

**Live Loading/Unloads**: A driver may deliver a load that must be discharged while the driver waits. Live unloads require the driver and the receiving clerk or supervisor to exchange paperwork and signatures to govern the unloading process.

**Longshoreman**: A person who loads and unloads goods, product, and cargo onto and off of a ship at a port.

**Lumpers**: Casual laborers hired on a case-by-case basis to assist drivers in the loading and unloading of product.

**Make-Bulk**: The act of combining multiple units of materials or finished goods to create a larger and more economical volume to ship.

**Manifest**: A document establishing the contents, origin, and destination of specified goods.

**Maquiladora**: A free trade zone designation bestowed on a manufacturing/assembly type plant that allows, for example, a plant located in a Mexican border town with the United States to receive imported materials or parts, add value through manufacturing or assembly processes and re-export the product while never entering the materials, components, or finished product into the Mexican marketplace for sale. Labor expertise and reduced costs may be achieved while Mexican duties and tariffs are waived because the product is re-exported.

**Marine Cargo Transit Warehouse**: A warehouse located at an ocean terminal to accommodate ocean borne freight transiting from ocean carrier to surface carrier.

**Master Production Schedule**: A plan detailing the production of specific items, quantities, and sequences spanning a defined time period, such as a weekly production schedule.

**Material Handling Equipment (MHE)**: Various manual and automated items of machinery used to move product from one location to another.

**Materials Requirements Plan**: An exploded documentation of materials and parts required to achieve the planned production schedule. Lead times for each material required indicate when to order, and a combination of materials on-hand, in-transit, and on-order are taken into consideration when establishing material volume to order.

**Merge-in-Transit**: Routing of a product's component parts, literature, and other component items that originate from various origins to a cross-dock or flow-through facility to be combined and deliver together. Bringing together components parts from several vendors to a cross-dock facility and bundling or packaging them together for single delivery to final customer.

**Metrics**: Quantifiable measures of performance established to assist in successfully managing various operations.

**Missed Shipment Ratio**: A common performance quality metric indicating the percentage of total shipments that were shipped on time: (Missed shipments) / (Total Shipments).

**Mixed Pallet**: Mixed items or stock-keeping units stacked on the same individual pallet.

**Moveable Bulkheads**: Physical separators that enable efficient and safe segregation of cargo for transport by various modes.

**Motor Carrier**: A vehicle mode for transporting product over the road.

**Motor Carrier Act of 1980**: The Motor Carrier Regulatory Reform and Modernization Act of 1980 was the determining act to mark the deregulation of motor carriers enabling greater rate-making and service-providing freedoms.

**National Association of Foreign-Trade Zones (NAFTZ.org)**: A not-for-profit organization with members having interest in U.S. Foreign-Trade Zones.

**National Association of Government Contractors (NAGC)**: A national trade association based in Washington, DC, for business owners engaged or interested in contracts with the government.

**National Motor Freight Classification (NMFC)**: Product groupings determined by the density of a product, the ease or difficulty of handling a product, the stowability or ease by which a product may efficiently be stored inside a trailer, and the liability/value of the freight. Instead of having to negotiate rates based on every type of product in existence, carriers and shippers may agree on a freight classification for the products shipped. Products with similar characteristics in the four primary factors will be thought of as equal when transporting. Product descriptions are grouped from freight classification 50 to 500 and may be placed within one of the NMFC categories for evaluating freight rating. The higher the freight classification, the higher the freight rate.

**Negotiation**: Discussions between two parties with the end goal of establishing a contract for services/goods.

**Non-Asset Based Freight Broker**: A third-party logistics (3PL) entity that negotiates with the carrier company on behalf of the warehouse.

**North American Industry Classification System (NAICS)**: The system used by business and government to classify business establishments according to the type of economic activity.

**Obsolescence**: Occurs when an object is no longer wanted or needed even though it may still be in working order.

**On-Time Delivery**: A Performance Ratio defined as (number of orders delivered on time / total number of orders delivered)

**On-Time Receiving**: A Performance Ratio defined as (number of inbounds received on time / total number of inbounds)

**On-Time Shipment**: A Performance Ratio defined as (number of orders delivered complete / total number of orders).

**Operating Ratio**: A measure of the carriers' ability to manage cost and revenue to achieve operating efficiencies. It is calculated as (Operating expenses / Operating revenue) x (100).

**Optimization Analysis**: Analysis undertaken to optimize the applicable process such as consolidation. This may be done manually or with the aid of software.

**Order Accuracy**: A quantifiable metric reflecting the accuracy of the order shipped against the actual items ordered.

**Order Cycle Time**: A Performance Ratio defined as (average time from order receipt to final delivery of order)

**Order Entry Accuracy**: A quantifiable metric measuring the accuracy of the order entered against the actual order received from the customer.

**Order Fill Rate**: Ratio of the number of orders shipped complete to that of the total number of orders placed (case fill rate = orders shipped complete / total number of orders placed). Used as a measure of customer service.

**Orders Shipped Complete**: A Performance Ratio defined as (number of orders shipped complete / total number of orders)

**Outbound Carrier**: The carrier transporting goods out of a distribution center.

**Outbound Order**: The order departing the warehouse.

**Outbound Shipping Clerk**: Verifies the count and condition of all outbound freight. Clerks work with picking operators, truck drivers, and inventory control personnel when necessary.

**Outbound Transportation Cost**: The sum of all costs associated with transporting goods out of a warehouse.

**Outsource**: Entrusting an outside party to manage an activity, process, or function of the firm. Typically, outsourcing agreements are made so that the work or service is performed by the most experienced party.

**Overage**: The situation in which stock on-hand exceeds the inventory levels plus stock required.

**Overages, Shortages, and Damages (OS/D)**: Overages, shortages, and damages pertaining to products within a warehouse or being transported.

**Overage, Shortages, and Damages (OS/D) Clerks**: The OS/D clerk manages product requiring special attention. It is the OS/D clerk's responsibility to manage the rework process and communicate with inventory control on the status of product in OS/D.

**Over the Road Driver**: The term used for the driver of the motor carrier.

**Pallet Jacks**: A tool used to lift and move pallets.

**Parallel Processing**: Processing multiple tasks simultaneously.

**Perfect Order**: An order delivered on time to the right place, with the right product, at the right quantities, to the right customer with the correct documentation.

**Performance Ratio**: Important metrics for evaluating performance. Ratio is the performance achieved expressed as a percentage of the total possible performance.

**Personnel Utilization**: A measure of the efficiency with which staff is used in the conduct of business.

**Physical Inventory**: Counting the entire warehouse contents. This provides for the reconciliation of errors that exist between the bookkeeping inventory records and the actual product inside the facility.

**Pickers**: Warehouse employees responsible for selecting and staging goods/items based on outbound orders scheduled for the designated time period.

**Picking**: Selecting and staging items based on the orders scheduled for the day's outbound shipping.

**Picking Cycle**: The frequency of the picking action. Part of the overall picking strategy.

**Picking Strategy/Strategies**: The approach determined optimal for selecting and staging goods/items within a warehouse for outbound shipping. These include area, zone, and bulk picking approaches.

**Picking Tunnel**: The automated picking from locations is conducted within a "tunnel" that traverses through the rack system.

**Pick-to-Light and Put-to-Light System**: An alternative to the A-Frame equipment technology, a pick-to-light system requires an order filler to move to the location of the SKU split case. Open cases align in horizontal rows, whereby, an order picker communicates to the picking function within the WMS that a specific order is being picked. After the operator affixes a barcode to a tote or carton for shipping, the scanned barcode initiates the picking of SKUs by piece. A light indicator illuminates at the location of the first item to be picked. A display near the light indicates the number of pieces to pull from the bin and place in the shipping tote. When the item is picked, the operator pushes a button to confirm, and the light turns off. The light indicator for the next SKU then illuminates, and the picker moves to the item. The process continues until the last light indicator displays a command that the order is complete and the operator can place the shipping tote on the conveyor for shipping.

**Pilferage**: The unauthorized consumption of or taking of product. Stealing.

**Point of Indifference**: The point where pursuing either path under consideration yields the identical results.

**Postponement**: To produce a product to a general stage but to delay the final commitment to customize the product according to actual orders received. Commitment of a product to a specific geographic market may also be delayed until the demand period for the product nears.

**Price Realism**: Price Realism can be an evaluation criteria element and is a measure of task understanding. It can prevent an entity from "buying in" to a project with a low price and then seeking to increase pricing after the contract is awarded. Price Realism also ensures that the bidder understands the contract requirements and that the price is realistic for the work performed.

**Price Reasonableness**: Price Reasonableness can be an evaluation criteria element to determine if a bid price is too high. This is a determination of what is a fair and reasonable price for performing the required services.

**Private Warehouse**: Private warehouses are operated by the company that owns the product being stored within the warehouse. The company may not own the building but does own all the products that come in and go out.

**Process Mapping**: An end-to-end layout of the designated process in an attempt to identify the granular aspects of the process for the purpose of optimizing each step and the overall process.

**Process Time**: The time required for the stipulated process to complete a cycle. This can become a key performance indicator (KPI).

**Process Variations**: Unplanned occurrences during a process that cause inefficiencies in the process and the end product.

**Product Assortment**: The variety of stock-keeping units in inventory to satisfy customer orders.

**Product Flow**: The path a product follows during a designated time period.

**Product Positioning Map**: The results of Analytical Hierarchy Processing (AHP) analysis can be placed within a Product Positioning map to illustrate the gaps between vendors on each selection criteria.

**Product Recall**: The formal instruction to locate previously distributed product and return it to a specified facility; sometimes returning to the origin shipper.

**Product Codes**: Unique identifying marks/numbers used to manage items in a warehouse. They may be imbedded into a radio frequency identification tag.

**Product Handling Personnel**: Product handling personnel manage the physical movement of product from the time of unloading to the ultimate loading of outbound orders.

**Product Staging**: The prepositioning of product into a designated area to facilitate efficient and expeditious movement.

**Production Changeover**: Time required to alter production equipment, materials, and processes to accommodate the production of a different item on the same line where various items are manufactured.

**Profit Margins**: The portion of product price that exceeds the cost of the product. Expressed as a percentage.

**Product Rotation**: The warehouse procedure for moving the oldest product out of the warehouse prior to shipping newer products on orders.

**Protective Packaging**: Various packaging techniques and applications intended to protect the contents from exposure and damage.

**Public Warehouse**: A public warehouse can be hired on a term basis and may store products from multiple customers. They can also be specialized warehouses that can accommodate unique goods or functions.

**Putaway**: The process of receiving and putting away stock in its final location in the warehouse.

**Quality Assurance/Quality Control (QA/QC) Plan**: A written plan detailing roles, responsibilities, processes, and standards to ensure the establishment and maintenance of process and product quality assurance and quality control.

**Radio Frequency Identification (RFID) Technology/Tags**: RFID tags use wireless radio-frequency technology to transfer data for the purpose of automatically identifying and tracking tags attached to objects. The tags contain electronically stored information.

**Receiving**: Accepting, unloading, and accounting for inbound freight arriving at a warehouse.

**Receiving Clerk**: Responsible for inspecting and counting all inbound freight. The clerks work with drivers to resolve discrepancies and when necessary provide instructions to drivers.

**Receiving Operator Performance Index (ROPI)**: A Key Performance Indicator (KPI) candidate. The ROPI is calculated using the standard receiving and putaway time and the employee's actual time for unloading and putting away product.

**Red/Green Light Trees**: Buildings equipped with red/green light trees protect warehouse employees by indicating safe or unsafe access conditions specific to a receiving or shipping doorway. Similar to a traffic light, with one red (stop) light and one green (go) light, each warehouse doorway would have a red/green light tree secured to the inside wall near the doorway and another red/green light tree secured to the outside wall near the doorway. A driver knows that the doorway is safe to back into or out of when the outside light is green. Red outside lights warn drivers to stop and don't enter or exit the doorway with your trailer. The inside red/green light tree is utilized in the same way so that when a red light appears by the dock door, a warehouse employee knows that it is unsafe to enter a trailer parked in the doorway.

**Relabeling**: Changing the product identification information affixed to the item; this may be done during the reverse logistics process to still gain benefit from the product.

**Replenishment Operators**: Warehouse employees responsible for replenishing designated storage/stockage areas.

**Request for Information (RFI)**: The purpose of an RFI is to solicit and collect information about the capabilities of various suppliers for a specific purpose. It is usually the first step in the process to attain contractual support. An RFI is usually followed by a request for proposal and a request for a quote.

**Request for Proposal (RFP)**: An RFP is a solicitation made to potential suppliers to inform them about the products and services wanted, the timeframe of the required support, and the evaluation criteria to be used in downselecting to a contract winner. The RFP requests that bidders submit proposals defining how they will provide the products/service requested.

**Request for Quote (RFQ)**: An RFQ is issued to request that interested bidders submit their price quote associated with providing the requested products or services.

**Required Delivery Date (RDD)**: The documented date that the customer requires the respective product to be provided.

**Reserve Product/Picking/Storage Area**: Reserve Product Areas contain product that is used to replenish active product areas.

**Reverse Logistics**: This refers to operations included in the reuse of products. Reverse logistics is the process of moving goods in the opposite direction from their planned destination for the purpose of capturing value or disposal. This can involve remanufacturing, refurbishing, and repair.

**Routing Costs**: The sum of costs involved in routing the product through the designated portion of the supply chain.

**Safety Stock**: Inventory held to protect against nonroutine events causing increased demand or interruptions in lead-time to receive replenishment stock at the time expected.

**Seasonal Fluctuations**: Product demand variations occurring in accordance with seasonal aspects such as Thanksgiving or summer.

**Seasonal Stock**: Inventory ordered and held in stock to fulfill demand initiated by the onset of a season, such as a holiday or other naturally occurring annual time period.

**Sensory-Based Cues**: Product cues alerting warehouse personnel to a defect and a potentially unsafe situation. All senses but taste are susceptible.

**Sequencing/Sequential Processing**: Organizing multiple shipment deliveries to a plant in a temporal order consistent with the manufacturing process requirements for the material/components at a specified time and location along the production line.

**Shipment Characteristics**: Items such as weight, cube, destination, ship and delivery dates, and other information critical for shipping and delivering the freight on time.

**Shipper Load and Count (SLC) Agreements**: The shipper of the product accepts responsibility for counting the product and releases the carrier from being responsible for the product accuracy.

**Shipping**: The transporting of orders from an origin to a destination. The term may also include the checking and loading of items contained on orders assigned, for example, to a specific trailer, container, or railcar.

**Shrink-Wrap Tunnel**: A heated tunnel that may be used with a conveyor system to securely apply protective film around a product.

**Shortage**: The situation in which the amount of product on-hand is less than the amount of product needed. This may be an indication of an operational process problem.

**Single-Double Lifts**: A single lift enables a one-pallet capacity move during putaway and picking. A double lift enables a two-pallet capacity move. (That is, one pallet in front of another and both are picked up and transported together by the double lift.) A double-reach truck compared to a single-reach truck has the capacity to reach two-deep in a product bay location.

**Single-Order Picking Strategy**: The picking strategy where a complete single order is picked and completed at one time.

**SKU**: Stock-Keeping-Unit. An individual product with its own unique alpha and/or numeric identifying code used for tracking.

**Slip-Sheets**: Thin pallet-size sheets that can be made of plastic, cardboard, or other substances and are used in commercial shipping.

**Sortation**: The act of sorting product. Sortation may be conducted manually, via mechanization, or automated.

**Space Utilization**: A cost and utilization measure. One example is defined as (total number of pallets in storage/total number of pallet positions available for storage).

**Special Handling**: The requirement for a product to be handled/moved with special consideration, which could include the use of general or tailored material handling equipment or environmental considerations.

**Specialized Warehouse**: Due the nature of varying products managed and services rendered by warehouses and distribution centers, all storage and distribution operations have elements of specialization. However, also standalone specialty warehouses can provide certain specific functions and handling and storing unique materials and goods. Examples include temperature controlled warehouses, import/export warehouses, and bonded warehouses.

**Speculative Stock**: Inventory held to meet anticipated but not routine demand. Marketing and sales, production, and inventory managers may have to speculate on the occurrence of many types of events. Some events may pertain to seasonal influences on demand, the anticipation of pricing or interest increases on the horizon, or even the worry of a shortage of transportation capacity availability.

**Stabilization**: Actions taken to eliminate variations in processes to enhance efficiencies, improve quality, and reduce costs.

**Staged Cargo**: Cargo that is moved to a designated location where it remains until it is moved to its final destination. This can be done as part of pre-issue preparation or due to safety or security concerns.

**Staging**: The process of preparing goods for use or transit; it may include relocating the goods to a designated staging area.

**Standardization**: The act of creating and implementing consistent procedural processes for the purpose of reducing variations, improving efficiency, improving quality, and reducing costs.

**Standard Putaway Time**: A time established as the amount required to perform receiving, unloading, and putaway of a specific product. Standard putaway time is used to calculate Receiving Operator Performance Index (ROPI), a Key Performance Indicator (KPI).

**Start-Up Costs**: The sum of all costs associated with starting a business; expenses incurred prior to actually beginning the conduct of business.

**Statement of Work (SOW)**: The document defining the specific work required; usually contained in a request for proposal (RFP).

**Stockout (Inventory Stockout)**: Not having enough product on-hand when the customer demands the product.

**Stockout Costs**: The sum of all costs associated with not meeting the demand for an item from current inventory. Costs include items such as lost sales/revenue, labor waste, and customer goodwill.

**Storage**: The holding of inventory within a warehouse in anticipation of future demand.

**Storage Agreements**: Bilateral agreements for one party to store items for another in exchange for payment; a legally binding contract for storage service.

**Stretch/Shrink Wrap**: The transparent film wrap around palletized cartons used to bind the items into a single unit.

**Sunk Cost**: A cost incurred in the past that cannot be recovered.

**Super Cargo**: The lead union clerk.

**Supplier Performance Index**: A quantitative measurement of the performance aspects of suppliers. A metric used to help manage suppliers; can also be used to evaluate competing suppliers.

**Supply Chain**: The composite "system" encompassing all activities involved in moving a product from the manufacturer to the customer.

**Supply Chain Management**: Management of the composite "system" encompassing all activities involved in moving a product from the manufacturer to the customer.

**Surcharge**: An extra charge beyond the contracted rate that is due to an unexpected influence causing the carriers cost to substantially increase. Medium-term fuel hikes that were unexpected at the time of contract negotiations may show up on the freight bill as a surcharge.

**Tariff**: A tariff is a monetary tax levied on imported goods. Contains the actual shipping rates that are based on the determined freight classification and are specific to a freight lane from origin to destination as well as the weight of the shipment.

**Task Interleaving**: Computer-assigned tasks that typically include varying work assignments according to the next critical task to be performed within the warehouse. Task assignment will take into account the location of the operator and the equipment under the operator's control.

**Tax Incentives**: Favorable tax allocations. States and municipalities may offer tax incentives to entice a company to select their area to locate or build its warehouse.

**Third-Party Logistics Provider (3PL)**: A third-party logistics provider of services within the supply chain that facilitates product and service exchanges between two other supply chain partners; typically a shipper/manufacturer and receiving customer.

**Throughput**: WMS systems have allowed for the assignment of various tasks, interwoven, to a single operator. A specific task will be assigned depending on critical factors that may include the location of the lift operator at the most recently completed task, type of equipment and skill level of the operator, time sensitivity of the outstanding tasks, and location and availability of other capable operators at time of task.

**Tie-High Configuration**: The tie is the configuration of a layer of product on the pallet that when every other layer is pivoted assists interlocking cartons together on a pallet.

**Time-Based Strategy**: A strategic plan consisting of activities and processes that are flexible and responsive to changes within the supply chain.

**Total Cost of Ownership**: All costs associated with ownership of an item or entity over the entire lifetime.

**Total Distribution Cost**: The sum of all costs associated with the distribution of a product from the manufacturer to the end user.

**Trailer on Flat Car (TOFC)**: An intermodal movement utilizing the flexibility of the motor carrier and the fuel efficiency of rail to reduce transportation costs and manage consistent service for customers.

**Transit Shed**: A warehouse facility attached to a merchant marine shipping dock that is used for temporary storage, processing, and cross-docking of export and import freight.

**Transit Time**: The time required by a carrier to transport an order from origin to destination.

**Transportation Costs**: The sum of all costs involved in transporting goods from manufacturer to customer.

**Transportation Management Systems (TMS)**: Transportation Management Systems (TMS) help to facilitate the shipping process. TMS can assist in optimizing the mode and carrier selection, along with optimizing load building configurations and even building consolidated loads from multiple individual orders.

**Transportation/Traffic Manager**: Responsible for hiring and managing the carrier base for customers. This function may be separate from or part of warehouse operations.

**Truckload (TL)**: Typically, one large shipment consisting or a single order or multiple orders transported from a single origin directly to the destination customer. At its most efficient, the orders fill up the maximum capacity available, thereby, weighting-out and cubing-out the entire trailer.

**Unitization**: Combining multiple pieces into single cartons, cases, or drums. Combining pieces inside a carton that allows warehouse operators to unitize individual cartons together, for example on a pallet, to create even greater economies of handling.

---

**Unitized Load**: Multiple cartons packed together on a pallet, typically stretch-wrapped together to create a single unit as opposed to handling and managing the cartons as individual shipping units.

**Upstream**: Looking or moving back up the supply chain toward the tier 1 suppliers, and then to the tier 2 suppliers, and on to tier 3 suppliers, and so forth. Moving toward the previous supply chain partner-in-line.

**U.S. Foreign-Trade Zones Board**: This provides information on applying for FTZ status, information about existing zone and subzones, and other information to help in the understanding of using or operating a U.S. FTZ.

**Utility Tractor Driver (UTR)**: A utility tractor driver.

**Value-Added Services (VAS)**: Unique customer service requirements that are performed by the warehouse and that add significant value for the customer.

**Variable Costs**: Variable costs are those that vary in proportion to output. Examples of variable costs are direct material costs and direct labor costs. As output increases, these costs increase; as output decreases, these costs decrease.

**Variances**: The difference between what is expected and what actually occurs. Process variances impact timeliness, quality, and cost.

**Walkie**: A low-lift pallet-jack type of material handling equipment.

**Warehouse**: A logistics facility designed to receive and store product to meet future customer demand. Value-added services may extend beyond long-term storage.

**Warehouse Management System (WMS)**: WMS is a suite of software that provides assistance in managing all aspects of warehouse operations with the goal of minimizing cost and maximizing performance.

**Warehousing Cost as a Percentage of Sales**: A Cost and Utilization Measure defined as (average warehousing cost per order / average order in terms of sales dollars).

**Warehousing Cost per Order**: A Cost and Utilization Measure defined as (total warehousing cost / total number of orders managed).

**Warehousing Education and Research Council (WERC)**: The professional association that focuses on distribution and warehouse management and its role in the supply chain.

**Warehousing Strategy**: The strategic selection of either public, private, or contract warehousing support.

**Warranty**: A contractual guarantee, containing enforceable provisions, which promised goods or services meet agreed upon expectations.

**Wheel Blocks**: Also called wheel chocks. These are traditionally wedges of hard material (wood, rubber, and metal) positioned behind a motor vehicle's wheels to prevent unintentional movement.

**Zone Picking Strategy**: Operators are assigned to designated inventory picking zones. Zone picking allows an operator to gain experience and familiarity in handling a finite set of SKUs and has been shown to improve picking accuracy and speed.

# Index

economies of sale, warehousing support, 3-5

electric rider lifts, 216

electronic information boards, 179

electronic picking tunnels, 48

eliminating variations in processes, 88

employees

assistance, 67

training options, 65

enterprise resource planning (ERP), 197

EOQ (economic order quantity) model, 129-130, 149

equipment

delivery routing models, 168

safe product movement, 177-179

safety devices, 177

utilization, 104-105

equipment

considerations for warehouse design, 51, 52

technology, 207

*AGVs (Automated Guided Vehicles), 212-213*

*AS/RS (Automated Storage and Retrieval Systems), 211-212*

*conveyor systems, 214-215*

*forklifts, 215-217*

*key terms, 217-219*

*piece picking automation, 213-214*

utilization, 61

ERP (enterprise resource planning), 197

errors and reconciliation, inventory management, 155

cycle counting, 155-156

physical inventory, 156-157

exchange of information, personnel, 66-67

exponential smoothing, 153

export licenses, 138

export warehouses, 222-225

EXW (ExWorks), 134

ExWorks (EXW), 134

# F

facilitation of product flow, distribution centers, 15

accumulation, sortation, allocation, and assortment, 16-17

cross-docking, 17-18

full-line stocking, 17

key terms, 19-20

postponement, 18-19

sequencing, 18

facility location analysis, 163-171

FAK (freight-all-kinds), 127

FAS (Free Alongside Ship), 134

fatalities among warehouse workers, 175

FCA (Free Carrier), 134

feedback, personnel, 67-68

fencing (physical security), 190

FIFO (first-in first-out) process, 47, 151

fire safety, 189

first-in first-out (FIFO) process, 47, 151

fixed costs analysis, warehouse selection, 36

floating warehouses, 227

flow-through distribution facility, 17

FOB (Free on Board) domestic transportation terms of sale, 132-134

shipping processes, 199
transportation processes, 199
vendor selection, 200-205
**worker productivity, 103-104**

# Z

**zone picking, 53, 93**
**zoning laws, 162**